SCIENCE
A
MEDITATIVE MIND

J. Richard Wingerter

University Press of America,® Inc.
Lanham · New York · Oxford

Copyright © 2003 by
University Press of America,® Inc.
4501 Forbes Boulevard
Suite 200
Lanham, Maryland 20706
UPA Acquisitions Department (301) 459-3366

PO Box 317
Oxford
OX2 9RU, UK

All rights reserved
Printed in the United States of America
British Library Cataloging in Publication Information Available

ISBN 0-7618-2546-0 (clothbound : alk. ppr.)
ISBN 0-7618-2547-9 (paperback : alk. ppr.)

∞™ The paper used in this publication meets the minimum
requirements of American National Standard for Information
Sciences—Permanence of Paper for Printed Library Materials,
ANSI Z39.48—1984

Contents

Acknowledgments	v
Introduction The Two Questions	ix
Chapter One Western Religion, Science, and the Meditative Mind	1
Chapter Two Eastern Religion, Science, and the Meditative Mind	57
Chapter Three Psychology: Science or Meditation?	89
Chapter Four Science and Silence	117
References	137
Index	155

Acknowledgments

For advice and for reading and commenting on the initial draft of this work, the author expresses thanks to Dr. Evelina M. Orteza y Miranda of the University of Calgary and R.E. Mark Lee of the Krishnamurti Foundation of America. Thanks is also expressed to Linda Lentz for preparation of the final copy for printing.

The author is most grateful and expresses thanks to the following copyright holders for permission to reprint material:

Daniel Menaker, "Werner Heisenberg: The Radical Thinker," *The New York Times Magazine*, October 17, 1999. Copyright 1999 by Daniel Menaker. Reprinted with the permission of Daniel Menaker.

The New York Times Magazine, "The Tricks Mirrors Play," by Margo Jefferson, and "I'm O.K. You're Selfish," by Andrew J. Cherlin, October 17, 1999. Copyright © 1999 by the New York Times Co. Reprinted with permission.

Richard Russo, "How 'I' Moved Heaven and Earth," *The New York Times Magazine*, October 17, 1999. Copyright 1999 by Richard Russo. Reprinted with the permission of Richard Russo.

Meditation is the emptying of the mind of all the things that the mind has put together. If you do that ... you will find that there is an extraordinary space in the mind, and that space is freedom.

The tricks of the mind are not meditation. Meditation is the beginning of self-knowledge, and without meditation, there is no self-knowledge.

J. Krishnamurti

Introduction

The Two Questions

In an earlier work[1] an attempt was made to provide a description of an observerless, choiceless looking at philosophy, specifically metaphysics, to see precisely and exactly what it was/is, and then of a going beyond it to that level of mind transcendent to the one on which metaphysics was/is created. In this work the same looking is described, but in this case the looking is at contemporary science and religion, again to see precisely or exactly what these are, and then again there is talk about a going beyond, this time beyond both science and religion rather than metaphysics, hence beyond a partially functioning mind and some of the things which it creates. This going beyond is to the level of the truly meditative mind, mind which functions totally and not partially, mind which is really silent and not merely made silent by a partially functioning mind.

There are two central questions which today call for serious attention on the part of those who are interested in and who study science and religion. In this work, both of these questions will be addressed, but always the basis or ground that is there when they are considered is that of the truly meditative mind. These two questions can be formulated as follows. Is it possible for a partially functioning mind, sooner or later, to acquire or to attain to a truly religious state, to become truly religious, as was suggested by all so-called great thinkers in the history of all religions, Eastern and Western, and which is also suggested by writers like some of those who will be mentioned later on in this book, writers who so often fashionably today speak of spirituality as different from, perhaps opposed to, religion in its traditional, ordinary, or usual sense? Secondly, is it possible for science, particularly when wedded to mysticism in the way that is suggested by some contemporary writers, eventually, even though not now, to fathom the mystery which is living, that is, ultimate reality, the unknown or unknowable, the immeasurable?

What is for the most part to be read hereafter is description of observerless looking at examples of things that have been written in relationship to the two questions just posed. Commentary on passages selected will be provided which will suggest that answers which might be and are given to these two questions are inadequate, limited as they are in every case to formulation by a merely partially functioning mind. The answers always bear the same limitations which are necessarily those of such a mind. Overall, it should be clear that answers to these questions are full and complete only when they come from out of a truly meditative mind.

What is provided here is not a critique of the passages cited but rather expression of what is seen when there is selfless, meditative mind observation of what is said in these cited passages, as well as the fullness of understanding which comes when there is such observation. Only this kind of observation in and of itself allows for that depth of insight which alone is a fully adequate answer to the two questions here posed.

Using figurative language one might say that what is presented in this book is a description of a walk taken by a truly meditative mind. As it walks about in the contemporary world, not, however, being party to its illusions, not sharing its false values, this mind observerlessly observes and listenerlessly listens, listens particularly to things said about this world, examples of which are given in the chapters of this book, and it takes note of activities carried on in this world, some of which are talked about in the quoted statements which follow and which are intended to bring sense and meaning to it. This walking, truly meditative mind realizes that in every case what is said is always out of a partially functioning mind and never out of a truly silent one. When, however, this truly meditative mind sees the limitations in what is said in relationship to this world and these activities, it goes beyond them to that level of mind to which alone can come deep understanding and profound wisdom. The meditative mind is ever mindful of the real dualism of time and the timeless, the limited and the unlimited, the measurable and the immeasurable as well as of the important implications of that realization. Without such realization, the mind functions to create the limited understanding, indeed the illusions which are cited in the quoted passages.

To put all of this somewhat differently, one can say that what is provided here in the commentary on passages is negation out of a truly quiet, silent, meditative mind. As Krishnamurti rightly said: "negative thinking ... is the highest comprehension."[2] Indeed, "to see the false as the false is the highest comprehension," and "only by understanding the false as the false is there freedom from it."[3] It is "awareness of the false as the false" which "is the freedom of truth."[4] Negation, however, when it is out of a truly meditative mind, is not denial or rejection of anything, but rather it is that full and complete understanding which are only of a truly meditative mind. Using a metaphor of William of Occam, who spoke out of a partially functioning mind when he talked about a razor which cuts away superfluous entities, one might say that when there is negation out of a truly meditative mind, such a mind cuts away illusions and falsehoods created by an inappropriately partially functioning mind, and then there is understanding which is full and complete, and this because it is out of a totally unconditioned, uncontaminated meditative mind. Such illusion and false belief are there, for example, whenever there is talk in cited passages which follow below about a real self, about the inner self or soul which is supposed or believed to be something real rather than merely a fashioning on the part of a partially functioning mind. Only a meditative mind fully sees the false in things said by a partially functioning mind about reality, which is the living. In such seeing, that is, there is seeing of the truth about things which are false, and this very seeing is itself seeing of truth which is living. As Krishnamurti said, "What is false must be put away if what is true is to be,"[5] if there is to be "the discovery of the true," which is "the understanding of the false."[6] Indeed, "seeing the false as the false, and the true as the true, is wisdom."[7]

Chapter One

Western Religion, Science, and the Meditative Mind

> A mind ... in ... isolation ... can never possibly understand what religion is. It can believe, it can have certain theories, concepts, formulas, it can try to identify itself with that which it calls God; but religion ... has nothing whatsoever to do with any belief, with any priest, with any church or so-called sacred book.
>
> The discovery of truth, or God ... demands great intelligence, which is not assertion of belief or disbelief, but the recognition of the hindrances created by lack of intelligence.
>
> J. Krishnamurti

When the meditative mind looks at Western religion, what it sees is a patchwork of various approaches, different paths which people follow in attempts by their partially functioning minds to arrive at what is a desired goal, called a Supreme Being, a sense of fulfillment in what is referred to as the ultimate, the unlimited, and so on. It is this same goal that fringe elements of the larger mainstream religions, like those referred to as New Age Religion, seek.

The focus in this chapter is on mainstream religion, specifically Christian. Examples from other Western religions, both mainstream and fringe, could have been cited to illustrate the same points.

The examples in this chapter relate to what might be called two dominant, mainstream approaches, namely, a traditional, left-brain approach which ever strives to harmonize faith and reason and a more right-brain, less strictly rational one, one which shows a penchant for using figurative rather than discursive language expression, one which therefore stresses not just the employment of reason or logic to express faith but which thinks that imagination, symbol, and metaphor are often better than ideas and concepts to talk about God, and about man in relationship to God.

The quotations in this chapter's first set are examples of left-brain, rational theism, and all of them are from Pope John Paul II's *Fides et Ratio*.[1] Those in the second set are from Anthony De Mello's *Awareness*,[2] which is an example of a less than strictly rational approach to religion. In the case of both sets of quotations, the commentary will make clear that all that is said by John Paul II and De Mello is partially functioning mind related and thus does very little by way of adequately answering the two questions mentioned in the introduction to this work. Something about full and complete answers to these questions will be said at points in this chapter and in the other chapters which follow. We turn more immediately to consideration of a first statement which is representative of Western theism.

1.0
John Paul II: Faith and Reason are like two wings on which the human spirit rises to the contemplation of truth, [and] belief is often humanly richer than mere evidence.

Commentator: The stances of those who reject religious belief as a way to truth, like those of the positivist and the atheist, as well as those of people who speak of such belief as important to their way of living, for example, theists and fideists, are all ideological, based upon some partially functioning mind derived conclusion. Beyond every ideological stance, religious or otherwise, is the truly meditative mind, and it alone communes with that which is profound and eternal, living truth, the ultimate, the immeasurable, life in all its rich beauty, but marred whenever a partially functioning mind casts its shadow on it, whenever anyone of its many activities militates against communion with it.

Several of the partially functioning mind's activities are the creation of vision(s) and speculation. Vision is always of the unreal. What is really beyond the limits of mind which partially functions is not anything envisioned but rather the fullness of understanding which comes to a totally functioning mind, a truly meditative mind.

There is no part or aspect of a partially functioning mind, be it reason or faith, which can fathom mystery, the unknown, what the theist, on the basis of a partially functioning mind, has concluded is reality and not illusion. Full and complete understanding of mystery or the unknowable unknown is only possible for a truly meditative mind and has nothing at all to do with belief or faith, neither with faith alone nor with a joining of faith and reason.

Even if it is related to some religious tradition and its scriptures,[3] vision is not related to the real, to what *is*. Any vision results from superficially functioning mind activity, and, since what this activity creates is merely temporal and relative, often false and illusory, there cannot be any depth and real insight related to it. One who prizes vision cannot say anything meaningful about life and freedom, and what is said by any visionary is mere projection from out of a partially functioning mind. Any vision described in so-called sacred writings is merely related to words which are of the past – memories, tradition, experience – to what is dead and not living. Thus with regard to it there cannot be depth of awareness and understanding. With regard to it there is only awareness which is consciousness and not that awareness which is not consciousness and related to a truly meditative mind. Such a mind is beyond and so not at all related to a futile, partially functioning mind related search for meaning in life.

Faith and reason are mere fragments of the mind, and what they give can at best be fragmentary and partial, if not false or illusory. The "believing soul" and the "faithful spirit" about which a theist readily speaks are not anything real but rather mere creations of a speculating, partially functioning mind.

The desire to know the truth about which theists speak is a futile desire for living truth which cannot be known. Living truth is of the unknown, the eternal, the immeasurable, and so it cannot be known in the sense that one knows some idea or concept for example. So-called fullness of truth about oneself, thought to be attainable for the theistically functioning mind, mind which functions in relation to ideas and concepts, is nothing else except that which a partially functioning mind fashions. Living truth can come to the mind only when there is that pure and untainted attention which is never there when there is faith or belief clouding the mind. The fullness of living truth comes only when there is choiceless and observerless observation, and not when the mind is engaged in contemplation, one of those activities which are very much prized by theists.

Neither reason functioning apart from faith nor reason acting as a handmaiden to a faith embodied in any so-called given revelation can give understanding which is related to living, to relating. Such understanding comes only to a quiet, silent, meditative mind. Any functioning of reason in relationship to living or to reality is engagement in futile, illusory activity.

The concern and esteem, moreover, which a theist has for reason or intellect is not real love for the mind of a theist is always partially functioning, which means self is always there, and, where there is self, there cannot be real love which is always and necessarily selfless. To desire, to reach for, and to seek to attain are all activities of the self, of the partially functioning mind, and living truth and total freedom of the mind never come, indeed cannot come to mind limited to awareness which is consciousness, to one which knows in relationship to such conscious awareness, to a mere fragment of the mind – reason, will, or some other.

To talk, as a theist might, about faith as capable of dispelling illusion and myth when it is brought into harmony with reason[4] is to voice preference for what might be called left-brain rather than right-brain thinking about religious matters. Even when one who favors such left-brain thinking talks about reason at the summit of thought, one is still talking about that which is limited, conditioned, and functioning in illusory fashion. The perception associated with reason, moreover, is always with perceiver present and never that perceiverless perception associated with a truly silent mind, that state of the truly meditative mind in which alone there can be direct communion with the transcendent, the absolute, the eternally alive. Reason, limited as it is, can never relate to the absolute. Any so-called meditation with meditator there cannot result in communion with that richness of the actually alive which touches a truly meditative mind.

What is sometimes called the light of reason and the theist's so-called "light of faith"[5] are not at all the light of the truly meditative mind. Only to the meditative mind come living truth, deepest wisdom, total freedom of the mind, the richness and depths of that which is living. Reason and faith, each alone, or in union with one another relate, not to the living but to the dead, and one who lives in relation to one or the other, or to both, lives in a charnel house – the past, memory, vision, tradition, experience, the temporal, the historical, and so on.

Reason is always limited as any truly critical philosopher knows, and as any truly meditative mind observerlessly sees. When reason rises to the God which the theist says is that revealed in scriptures of some kind, and which is said by theists to be identical to the Supreme Being of classical philosophy, it is dealing with mere creations of a partially functioning mind, and so it is involved in illusory, partially functioning mind movement or activity. Faith, as much as reason, is merely thought, and thought, no matter what its form, is related to the dead and not to the new and the living, has to do with the part and not the whole, with the past and the future as mere projection out of the past, with tradition, experience, the recognized.

1.1
John Paul II: As a work of critical reason in the light of faith, theology presupposes and requires in all its research a reason formed and educated to concept and argument.

Commentator: If theology and philosophy merely presuppose and require "a reason formed and educated to concept and argument," then both theology and philosophy are ever limited to the level of a merely partially functioning mind, to a mere fragment of the mind, be it reason or will. Beyond a partially functioning mind, mind functioning in terms of reason or will, hence beyond theology and philosophy as these have been and still are defined, beyond, that is, concept and argument, is the truly meditative mind. Were philosophy at a deeper, more profound level to be love of wisdom, wisdom related to actual living and relating, then it would be, at least sometimes, of the level of the truly silent mind and not merely and always of a merely partially functioning mind level.

Any philosophy related to reason joined to faith, making possible, at the highest levels of speculative thinking, the attainment of so-called profound truth and understanding[6] is actually as limited as is any philosophy based upon autonomous reason, reason functioning apart from any faith. Any theism which weds faith to reason, any skeptical agnosticism, any fideism which would altogether deny faith has any relationship to reason, indeed any system of thought based upon some specific premise or conclusion, are all partially functioning mind derived ideological stances, and no one of these relates to the depth of understanding, insight, and intelligence which are possible when there is the state of the truly meditative mind.

It is traditional and not mainstream contemporary philosophy which prizes speculation, and which would claim that speculation relates to sufficiently rigorous and critical thinking.[7] Though the results of speculation might be called a logically coherent body of knowledge by theists, these are not anything real nor are they at all related to the real. Just as unreal as this so-called body of knowledge is the speculator, that is, the self, the partially functioning mind which produces this so-called body of knowledge. There may be unity of the content of this so-called body of knowledge, but such unity is not actual, real, not the wholeness with which a truly meditative mind communes, life in all its dynamism and richness.

It is only a theistic, traditionally or classically oriented, pre-critical philosopher that would talk about the value of speculative knowledge. Such so-called knowledge is merely a creation of a partially functioning mind. Speculation, contemporary philosophers have shown and a truly meditative mind sees, choicelessly and observerlessly, relates not to anything real but only to what is imagined, the unreal rather than the real.

Contemporary analytic philosophy is not of one of the schools of philosophy about which the traditional philosopher sometimes speaks.[8] Unlike classical metaphysicians, analytic philosophers do not characterize reason as that which can intuit the first principles of being and which can draw appropriate conclusions from these intuited principles.[9] Though these conclusions might be made logically coherent, they relate, not to what is actual, real, that which is living but rather to what is unreal, merely abstract fashionings of a partially functioning mind.

The analysis, judgment-making, and evaluation associated with the exercise of reason,[10] and with or without faith joined to it, are of a limited, partially functioning mind. Deepest understanding cannot come from any such partially functioning mind activity. Understanding in depth comes not from any profession of faith in a merely mind-projected God supposedly at work in the world and the events of history[11] but rather comes out of choiceless and observerless attention to what *is*, out of the understanding which comes of such attention.

The principle of non-contradiction, for example, or that of causality so often invoked by theists,[12] has value, but this value is as limited as is the partially functioning mind to which both these principles relate. This mind and these principles are helpful when the concern is the scientific or the technological, for example, but, when the concern is the living, what is appropriate is not the partially functioning mind and its principles but rather the truly meditative mind, mind on a level truly transcendent to the level of mind on which there is partial mind functioning, really transcendent to the principles which relate to this level of the mind. Post-Kantian, contemporary philosophy has shown, and a truly quiet, meditative mind sees, choicelessly and observerlessly, that reason cannot attain the deepest good and living truth.[13] To say, as Christian theists do, that this good and this truth are incarnated in Jesus Christ is to indicate that one has drawn a particular partially functioning mind related conclusion, a conclusion merely opposite to another partially functioning mind drawn one which is a denial that this is so, related as it is to some organized religion other than the Christian or to some areligious or antireligious ideological stance, for example, agnosticism or atheism. When, as a result of deliberation and assessment, one chooses to side with anyone of these conclusions rather than another, one is merely on the level of mind which functions in terms of opposites, and to such a mind living truth never comes.

Faith is the result of mind activity, and any and all such activity is of the limited, the temporal. Any and all knowledge relates to that level of the mind which is partial and fragmentary, and how can that which is partial or fragmentary possibly know the whole? It cannot. There can, however, be awareness of the whole or the eternal but only when mind is in a totally quiet state.

Faith, even if distinguished from what the theist refers to as philosophical knowledge, is merely a creation of or a fashioning by a partially functioning mind. Intellect is partial and fragmentary, but so is faith, and thus, just like intellect or reason, it cannot give the mind fullness of living truth. Philosophy, like any science, is and always has been partially functioning mind related, and, since faith is also merely so related, it is only a truly meditative mind and not faith which relates to the whole, to living truth in its fullness, to the eternal beyond the merely temporal, to the immeasurable beyond the measurable.

The planes which faith informed by reason is said to be able to reach are merely illusory fashionings by a partially functioning mind. Such planes are merely the products of imagination, hence nothing real, and not related to the real, to what *is*, but merely to what might be wished for, longed for. To live a life based upon a union of faith and reason is not to live with mind totally free, completely unconditioned, absolutely free of the shackles theists might refer to as the principles and norms of a faith served by reason.[14] One who lives out of such principles and norms is living in accord with a particular form of a partially functioning mind. Beyond these principles and norms, and beyond any other partially functioning mind created principles and norms is the truly meditative mind, that mind which is truly and deeply moral. The intelligence of this mind is not merely abstract or verbal but rather living. Only such living, loving intelligence, related as it is to total goodness and absolute order, and beyond any so-called moral good and its opposite, a so-called evil, can make for living which is holistic and profoundly human.

To engage reason, conscience, in relationship to the making of ethical decisions, as the theist would have us do,[15] is to engage a merely partially functioning mind and not at all for mind to be in a truly meditative state, a state of mind which relates to the real, the what *is*. Such a state of mind has nothing at all to do with some vision or other, something unreal, merely imagined, wished for, longed for, nothing at all to do with that seeking for a goal which is merely escaping from what *is* rather than an understanding of what *is* and a going beyond it when this is appropriate. To seek for a goal is to follow a path, but there is no path which leads to living truth. Any synthesis of faith and reason, of philosophy and theology is merely something mind created even if called a unity of knowledge, and any speculative thinking and its products, even if called profound and insightful, relate to partially functioning mind activity which fashions illusions.

Knowledge of living truth cannot be produced in the way that theists suggest, nor in any other way. Indeed, knowledge of truth which is living is an impossibility. There can be, however, awareness without consciousness of what living truth is, that is, communion with the truth which is dynamic and ever-changing, the living, the really deep and beautiful.

In its long history, philosophy has engaged the minds of people who searched for and found answers to questions about the meaning of life, but in every case the answers were false for they were fashioned by a partially functioning mind, related not to the real but only to what was imagined, the unreal rather than the real. An answer to the question of the meaning of life is given only to a truly quiet, meditative mind for only such a mind is properly disposed to receive the answer given to it by living reality, by life itself. Fullness of humanity, moreover, relates not just to rational functioning of the mind but also and more significantly to quietness of the mind, real quietness, however, and not a partially functioning mind created or imposed quietness.

What often today is called human activity, whether or not related to the exercise of reason, and whether or not this reason is wedded to faith is not the loving action of mind in a truly meditative state. It is merely related to the functioning of a superficial mind, and that which is superficial cannot ever move beyond the merely superficial. For depth of understanding and of living in accord with deep understanding there must be the state of the meditative mind. Only mind in such a state can answer questions related to living, to the meaning and significance of living.

Past and present philosophers have spent, and still spend a great deal of time talking about the objective and the subjective, the external and the internal, or about culture, civilization, or society and the individual. Objective and subjective, however, are fashionings of a partially functioning mind, that functioning which is dualistic, involving separation of the observer from the observed, the knower from the known, hence not at all related to what is living, life in all its richness and vibrancy. To live in relationship to what is objective or subjective, or to a partially functioning mind attempt to unite the subjective and the objective, is to live in relationship not to the living but rather to what is inert, dead. So called truths deriving from what is called objective reality are not living truth but rather mere projections out of a mind which functions in relationship to a mere fragment of the mind – reason, reflection, and so on. Such projections of a functioning mind, like the theist's personal God and the human person as spiritual being,[16] are not related to what *is* – what is real. These and similar projections of a partially functioning mind are merely speculative, products of the imagination, and not at all related to what *is* but rather to what is unreal – the ideal, the supposed, the envisioned, the imagined, the what ought to be, could be, should be, that which is merely believed to be.

When reason's scope is broadened in the way that a theist like John Paul II suggests,[17] when it is permitted to go beyond the limits which many, if not most, philosophers today impose upon it, any post-Kantian, hence post-critical thinker would say that this reflects a failure to respect the proper bounds or limits of reason. Then, of course, there is the fideist who, in his recourse to faith alone when the matter is anything to do with religion, does not at all attempt to broaden reason's scope nor, indeed, to show or demonstrate any respect for reason and its set limits. As matters stand, whatever any theist, fideist, or contemporary philosopher who is mindful of the Kantian critique of reason, or who in some fashion or to some significant degree adheres to this critique of reason or of knowledge might say are the proper place and bounds of reason is only partially functioning mind related. Never in anything which anyone of them says is there any indication of the importance of truly silent mind awareness, awareness which allows reason its proper place. When there is really such awareness, there is also an understanding of when there should be a cessation of reasoning, not however so that it might merely be replaced by fideistic faith but rather so that there is just the truly meditative mind and no partially functioning mind activity at all. When the mind functions inappropriately, that is partially in areas where and at times when it should not, then it is merely projecting illusion or falsehood out of itself. Such inappropriateness is there, for example, when, in attempts to know and understand reality or the living, it speculates, imagines, envisions, entertains what could be or what should be, and so on. Only when all such projecting activity of a partially functioning mind ceases, is there that fullness of awareness and understanding which relate, not to awareness which is consciousness, to reason, knowledge, a partially functioning mind but rather to awareness other than consciousness, to mind which is really quiet, silent, meditative.

1.2
John Paul II: The appeal to tradition is not a mere remembrance of the past; it involves rather the recognition of a cultural heritage which belongs to all of humanity .... Precisely by being rooted in the tradition will we be able today to develop for the future an original, new, and constructive mode of thinking.
Commentator: Traditions and cultures are of the past since they relate to the recognized, the experienced, the known, and not at all to the truly new, the unknown, the living. To belong to a tradition is to live in a house of the dead even when there is a partially functioning mind projecting a future supposedly, but not really, beyond the past. A so-called new and constructive mode of thinking is not really new for all thinking is past-related, related to memory, the known, the experienced, in a word to the old and the dead and not to the real and the living, the eternally alive, that which is dynamic and ever-changing rather than inert and lifeless.

Any tradition – theological, philosophical, or some other – is of the past since it is related to some aspect of a partially functioning mind – reason, will, and so on – related, that is, to the past, memory, time, the relative. It cannot, therefore, be dynamic. Only the living is dynamic, and so any thought and any tradition which is thought-related are inert, dead. Any philosophy of being and any contemporary philosophy of the analytic or European continental type cannot directly relate to the new, to reality. Only a truly meditative mind is fully open to reality.

When a philosopher of being, employing a partially functioning mind, strives to reach fulfillment in what is called God, this is an instance of the known striving futilely to know the unknown, the observer separate from the observed projecting from out of itself a merely mind-created, purely imaginary something characterized as God, the eternal.

Faith and so-called metaphysical reasoning are both aspects of a partially functioning mind, functioning without due respect for the proper limits of reason, of logic, and neither can give the fullness of understanding and meaning which come only to a meditative mind. Both faith and reason have to do with the limits of a partially functioning mind and not with the meditative mind and the ultimate goodness with which such a mind communes. Such goodness is of the real or the living and not of the merely abstract. It is not that which is called good by a society or religion, a good opposed to what that same society or religion calls evil.

Moral theology and philosophical ethics are totally unrelated to living truth, the ultimate good, love without self, total freedom of mind completely unconditioned, that is mind not conditioned by society, by culture, or by religion on the one hand, or by some counter-societal, antireligious, or counter-cultural group on the other, indeed not even by one's own so-called personal experiences. Ultimate goodness, real goodness, and not goodness as determined by a partially functioning mind is beyond subjective and utilitarian ethics but also beyond an ethics and morality associated with a philosophy of being.

Any clarity which comes of philosophical inquiry in the usual or ordinary sense can only be verbal and intellectual in kind, and any relationship established as a result of such an inquiry between so-called truth and life is of the unreal, the merely mind projected, not of the real. Abstract, intellectual truth, as well as any language which expresses such truth or so-called truth are related to the known, the experienced, the recognized, and as such they cannot relate actually or directly to life, the transcendent, living truth. To think that they can is to think that the known can know the unknown, the measurable can measure the immeasurable, that, by means of the relative and the temporal, one can go to the absolute, the eternal. The relative and the temporal, and things related to the relative and the temporal, like intellectual truth or so-called truth, dubbed doctrinal or otherwise, must cease, must be negated in the awareness of the meditative mind before there can be the living, the transcendent, the eternal. Such awareness is not at all related to so-called understanding that is said by theists to come from faith.

Faith, since it is necessarily related to a partially functioning mind, has nothing to do with that which is deep. It is always related not only to the superficial but to the illusory, the merely imagined, the merely mind-fashioned, the what should be or ought to be, to the merely wished for or longed for but never to reality, to what *is* and that which is really beyond what *is* when the what *is* is that which makes for confusion and chaos in living, in relating.

There is no authority which and no leader who can lead people to a deep understanding of life. Such understanding does not come from a partially functioning mind projection called truth but rather is discovery of truth when there is the state of the meditative mind which has to do with choiceless and observerless observation of what *is* and not with what is imagined, or envisioned, or projected as image or concept. The state of the meditative mind does not relate to knowledge in the sense of accumulations of images and concepts by an aspect of a partially functioning mind – reason, reflection, imagination – but relates to self-knowing, knowing what the self is and does, that it is not anything real but something merely mind-fashioned and so often responsible for bringing about confusion and chaos in living.

Philosophy as it is and has been is not and cannot be deep wisdom. True wisdom relates not to a partially functioning mind, this is to say, but only to a truly meditative one, one which understands completely and whose insights are total, one which alone communes with the really true, the really good, the really beautiful in their actuality and not as the mere abstractions which the Christian theist's partially functioning mind first of all fashions and then makes identical to what he says is the Christ as revealed. When Christian philosophers and theologians advise people to entrust themselves to what they call the revealed Christ, they are merely engaging their partially functioning minds, and whatever is of the partial or the fragmentary cannot be anything other than partial and fragmentary and very often is altogether illusory.

Reflection, one activity of a partially functioning mind, cannot relate to what is deep, the living in all its dynamism and resplendency, cannot at all attain to the fullness of truth but rather only to that which is superficial and relative. The fullness of truth can never be grasped by a partially functioning mind, cannot at all be known, but of it there can be awareness without consciousness when there is the silence of the meditative mind.

So-called Christian truth offered to people by Christian theists and other so-called truth which is offered by any other Western or Eastern religion as an answer to man's psychological problems are all partially functioning mind fashioned conclusions. Beyond any so-called truth and so-called answers is the fullness of understanding which relates to a truly silent mind, mind beyond all reason and/or faith derived conclusions. Only such understanding is a full answer to people's psychological problems, to their problems of living. Only the loving intelligence and integral wisdom of a truly meditative mind make for depth of understanding and the completely different way of living which such understanding makes possible, a way of living which alone can be deeply meaningful, filled with real joy and love which is eternal. This deeply moral way of living is total freedom of the mind, with right relationship to everyone and to everything.

1.3
John Paul II: In both East and West we may trace a journey which has led humanity down the centuries to meet and engage truth more and more deeply. It is a journey which has unfolded – as it must – within the horizon of personal self-consciousness: The more human beings know reality and the world, the more they know themselves in their uniqueness.

Commentator: This journey is that of a partially functioning mind, of left rather than right-brain thinking, and it constitutes greater and greater conditioning of the mind, more and more prizing of the unreal, indeed of the false, rather than of living truth.

Person is not anything real, and though, as a mere creation out of thought, of a partially functioning mind, it does relate to consciousness of self, it is that which prevents deep understanding and insight. Human beings cannot know reality, and knowing themselves in relationship to what is called their unique individuality rather than their being aware of the nothingness, the unreality, hence the mere ideality which is the self, person, or individual relates not to living truth but to something merely fashioned by a partially functioning mind, that is, an image or a concept. The concept of person as a so-called free and intelligent subject is a creation of a partially functioning mind. Not only is a concept not real, merely an abstraction from what is real, what *is*, but person likewise is not anything real but merely a fashioning by a partially functioning mind. Thus, since that which is unreal could not possibly know that which is real, it is impossible for a person to know God, truth, and goodness. Indeed, God, living truth, and real goodness, that is, the eternal or the immeasurable cannot be known at all in the ordinary sense of knowing, that is, by means of anything which is time related – experience, images, concepts, and so on. Though these cannot be known, however, there can be awareness of them when there is a truly silent, meditative mind.

The mystery of life cannot be known by a partially functioning mind. This is to say that it cannot be imagined, envisioned, conceptualized, imaged, and so on. There is quiet mind awareness of mystery and the understanding which such awareness brings, but these are possible only when there is self-knowledge rather than a partially functioning mind so-called knowing of a supposedly real self which is not really different from the ego. Genuine self-knowledge is a moment to moment knowing not at all related to any accumulation or gathering on the part of a partially functioning mind, a moment to moment knowing of what the self really is and what it does, even foolishly imagining that it can penetrate mystery, can know the unknowable unknown. That which a partially functioning mind thinks it can attain in faith is mere projection from out of its own partial, fragmentary, even illusory, functioning. Such a mind-projected or mind-made mystery is not anything real, not that real mystery with which only a truly meditative mind can commune.

Mind ceaselessly striving or making effort to experience communion with mystery is a merely partially functioning mind, mind which cannot, in its never ceasing activity, commune with living mystery, the vastness and depth which is life itself. No matter how far reason extends its range never by its very nature can it touch mystery. Mystery is real, however, and there can be communion with it when there is the state of the really silent mind but not when there is merely some fragment of the mind, like reason or will, striving for communion with it.

Knowledge cannot at all relate to the unity or wholeness with which a truly silent mind communes. Any so-called synthesis or wholeness about which a theist speaks is mere projection from out of a partially functioning mind, and the partial or what is related to the partial cannot yield the whole. The eternal, that which is really the whole, is not opposite to time but rather is deeper than time, envelops time. Only a deep mind, that is, a meditative mind can commune with the deep. The deep, moreover, is neither exoteric nor esoteric. That which is interior is merely the opposite of what is exterior, something therefore merely mind projected, and always to think and talk in terms of opposites is always to walk in the hallway of opposites, unaware that there is anything beyond this hallway. Fullness of living truth and meaning are possible only in the state of the truly silent, meditative mind.

Logic is not that which can open the door to the beyond, the absolute, the immeasurable, that beyond which is not at all related to conditioning of the mind by culture, by religion, by society, by one's own so-called personal experiences. When logic is ended, when there is a stop to the functioning of that fragment of the mind to which logic relates, and this by the total attention of a truly meditative mind, then there can be communion with mystery, mystery, however, which is actual, real, of the living, and not mystery as created by or as projection out of a partially functioning mind. Also mere projection by a partially functioning mind, relating to the unreal, the false, or the illusory are the theist's so-called "essence of God" and essence of man.[18] Such projections of a partially functioning mind are indications of a failure on the part of theists to respect the proper limits of reason.

That there is no path of any kind to living truth is something which a truly meditative mind sees and understands, and thus what it also sees and understands is that so-called revelation and faith in it cannot successfully aid reason in any futile attempt to know and understand mystery. Effort of any kind related to functioning on the part of reason, will, faith, imagination, or any other fragment of the mind cannot bring real understanding of life, of mystery, of the unknown to the mind. Autonomous reason or reason dependent on something other than itself, for example, faith, cannot penetrate mystery. Living truth is not contained in any words, signs, or symbols, no matter the source, be it so-called revelation or some other, but it is there for discovery when the mind is totally silent, truly meditative, mind beyond all partial functioning and any and all products of such functioning, including culture, tradition, revelation, and so on.

Fullness of light, of understanding, that is, comes only to a truly meditative mind, not to a mind tied to some conclusion arrived at because of partial mind functioning, be it a Christian one or some other one which is equally and necessarily divisive, causing confusion and conflict. So-called believers and nonbelievers are always on an exclusive partially functioning mind level, and, when there is only this level of the mind, there cannot be holistic insight and understanding, possible only when there is observation without observer there, be this observer a so-called believer or nonbeliever, one who merely holds to a particular conviction of one kind or another. Faith is fragmentary, thought is fragmentary, and thus neither nor both can relate to what is deep and authentic. Only the meditative mind communes with the living rather than with the merely abstract, the merely imagined, what is unreal, all of which are old and superficial rather than new and deep.

1.4
John Paul II: One thing is certain: Attention to the spiritual journey of ... masters can only give greater momentum to both the search for truth and the effort to apply the results of that search to the service of humanity.

Commentator: Attention here is not of a meditative mind for here there is an attender. A so-called master on a spiritual journey is a self, a partially functioning mind seeking to fulfill itself, enhance itself. Living truth cannot be sought out by a partially functioning mind for it comes only to a totally silent mind. Effort to apply the results of a search for the living or the eternal is effort merely to apply something projected out of or imagined by a partially functioning mind, to make prevail what one thinks ought to be or should be, that which cannot be anything other than unreal.

There is no path to living truth, hence no final goal to be reached by a so-called higher self or soul merely supposed to be real but not really so. Any such goal is never anything other than something which is fashioned by a partially and improperly functioning mind which creates illusions, no matter how this goal is characterized, in Christian or other terms.

One can say that there is a path to the understanding of scientific truths which are intellectual in kind, the path of reason and evidence-gathering. When, however, theists employ speculation which has no basis in fact or in what *is* but only in what might be wished for, imagined, and so on, and when fideists resort to belief or faith against reason if they deem this necessary, all of them are not on a path to truth but rather on one to illusion. The mind of neither the theist nor that of the fideist is the truly religious mind. Beyond theism and fideism, and also beyond a partially, but properly functioning mathematical, scientific, or practical mind is the truly meditative mind, the only mind or level of the mind which relates to living truth.

The answer to the question of the meaning of life is not anything which can be sought out by a partially functioning mind. Human reason, even at its so-called highest levels, is limited to the temporal and the relative, and so it cannot provide depth of understanding and insight into living truth. The so-called human person who reasons is not anything real, and any wish or desire on the part of this so-called person to grasp what is called truth is likewise merely related to the unreal.

Living truth comes not to one who wishes and wants, seeks and strives, but only to a mind which is silent. Engaging oneself in some so-called personal search for God or life's meaning will eventually result in finding or discovery, but what is found or discovered will be, indeed could only be, projection from out of a partially functioning mind. Such a mind projection, since it is something unreal, is illusory when thought to be something real, and, when mind is tied to illusion, it is not free but rather enslaved to the results of partial mind functioning.

Living truth comes only to a truly silent mind, not to a mind which strives to go beyond the point at which it now finds itself, not to a mind engaged in ceaseless activity to get, to acquire, to achieve, to fulfill itself. Reason on a so-called path to the absolute which a theist might say is proper because outlined by some so-called spiritual master moves in illusory fashion, and where it arrives in the end can never be anything real or actual but merely what is imagined, ideated, unreal.

There are so-called spiritual masters associated with so-called New Age religion or with places where so-called consciousness-raising seminars and workshops are conducted. All of these so-called spiritual masters are supposedly people who, by effort, concentration, focusing, disciplining of the mind, have discovered their so-called inner, higher, deeper, real selves. What they have really done, however, is merely enhanced or aggrandized merely thought or partially functioning mind projections labeled or called their higher or deeper selves, their souls, their spiritual essences, all of which are not fundamentally different from what they call the lower selves or egos. Self, no matter what its form, is always something merely mind-created, thought-fashioned.

Because of their supposed deep level of awareness and understanding, these so-called spiritual masters present themselves as qualified to assist others, as they might put it, to deepen their connection to so-called spirit, make a path which leads to joy and well-being, find greater connection with self, thought to be but not really real, attain to clarity of mind, come to a recognition of their own inner essence, and so on, all of which relate to mere fashionings of a partially functioning mind which are illusions because they are thought to be real when actually they are not.

Besides these so-called spiritual masters there are others representative of long-standing religious and cultural traditions. Fundamentally, however, there is no difference between the more contemporary and the older type of spiritual master for in the case of both types there is the same deflating of all awareness to consciousness, limitation of the mind to only its partially functioning aspects, and imagining that there is an inner, deeper, higher self which is real, a soul or spiritual essence.

As it is, all cultures and subcultures, and all mainstream and fringe religions which produced/produce so-called spiritual masters have been/are totally related to a partially functioning mind, have been/are experience-related, related to memory, the past, tradition, and so on, in a word to what is dead. Even what is referred to as a "living tradition" is of the house of the dead. Those who live the tradition are living in relationship to the dead – the past, memory, and so on – and not at all to the living, with real openness to the transcendent, the eternally alive, life in all its abundant richness.

Spiritual masters often talk about different paths to truth. However, though there can be paths to what partially functioning minds project as truth, one cannot really and rightfully say there are various paths to living truth which one can take. Any values related to what partially functioning minds project and call truth are not real values but merely creations or fashionings of a partially functioning mind.

Life is dynamic and of the eternal now whereas cultures and religions, and the experiences to which they relate, are of the past, that is, are based upon and derived from tradition of some kind. They therefore are of what is dead rather than of the living and the dynamic. Any so-called new approaches and ways to truth[19] are not really of the new or the unknown since they are of the past, of tradition, of experiences, thus of what is dead rather than living. Cultures and religions never were and never are open to the really new. The so-called new experiences which any given culture or religion might attempt to make its own are not really new but rather are partially functioning mind projections of what is old, dead.

1.5
John Paul II: Men and women may ... come to the fullness of truth about themselves.

Commentator: One cannot arrive at living truth, but living truth can come to a mind that is meditative, silently receptive. Partial, fragmentary truths of an abstract kind are possible in relationship to mathematics, science, and everyday practical matters. A theist's so-called philosophical truths arrived at by means of speculation, however, are neither really abstract truths nor living truth, and the same is true as regards so-called truths of faith or of revelation, whether these are said to be susceptible to harmonization with philosophical reasoning as in the case of theism or sometimes not at all so as in the case of fideism. That which is of the dead – tradition, religious or otherwise – cannot relate to truth which is living.

Beyond the limits of reason and the partially functioning mind to which it relates is the truly silent mind. Though the latter serves no pragmatic, utilitarian, temporal, historical, and futuristically related purpose, it nevertheless is of immense and immeasurable worth or value. When primacy of place is given to the truly silent mind, then fullness of intelligence, insight, understanding, and wisdom can be there in fully functioning minds, giving a depth of meaning and significance to living which are not there now in this world which is ours.

Reason by its very nature, limited and fragmentary as it is, cannot grasp truth which is living even if it is said by theists to be capable of focusing itself on truth in its totality. Reason properly relates to mathematical, scientific, and practical truths which are necessarily abstract but not at all to the fullness of living truth which comes only to a profoundly silent, truly religious mind, and not to a mind which is merely functioning partially, engaging itself in activity of some kind.

There is no mystery of one's own personal life.[20] There is mystery, but it is neither personal nor impersonal. It simply is, and with it there can be communion, but only when the mind is truly silent and not when the mind is partially functioning. There is absolute truth which is living, but, because it is living, it is not related to what the theist calls revelation, that which is of the dead – the past, experiences, tradition, words, that is, scriptures, so-called sacred writings, all of which prevent real openness to the transcendent. Indeed, there cannot be openness to the transcendent when self is there, whether or not it is said to be a free and autonomous creature or creation of God.[21] Creature is the opposite of creator, and autonomy the opposite of dependency. Where there are opposites, there is self and not the truly meditative state of mind beyond self and all opposites.

Ancient philosophy was fragmentary when it focused attention on God, or rather on what it projected God to be, and modern philosophy was fragmentary also when it focused attention upon man. The attention which was focused in both cases was not the attention of a truly meditative mind, holistic and integral as it is but rather was that of a partially functioning mind.

The theist who talks about human reason desiring or longing to know and to know more profoundly[22] disregards the Kantian critique of reason. It also disregards contemporary analytic philosophy's restriction of reason to only that with which it can properly deal.

There is truth which is living, and which transcends the subjectivity and the objectivity which together make up a set of opposites. Living truth is beyond all opposites, and communion with it is possible only when there is the state of mind which does not function in terms of opposites, indeed which does not function partially at all, or which functions totally and not partially as it ordinarily does. Reason which would rise to the so-called summit of thought,[23] which would rise to the theist's so-called truth of being, is reason which does not recognize its proper limits. There are depths of living and meaning beyond reason, there is living truth to which reason cannot relate, but these are there only when there is the state of mind beyond that wherein reason operates, that is, the state of the totally silent, meditative mind.

There is no path to living truth, not even that of time and history. Living truth is of the eternally alive, what is dynamic and ever-changing, whereas time and history relate to what is dead – time as past and future, and present merely as a way from the past to the future. Related to the dead are tradition and experience, including scripturally described or related so-called revelation, and to live in relationship to some mind-projected goal, something of the imagination, hence of the unreal, the what should be rather than the what *is* is to live with the dead, the stagnant.

Theists often inveigh against what they call immanentism,[24] a failure, they would say, to give proper recognition to what they call the transcendent. Immanence and transcendence, however, constitute a false dualism, a mere set of opposites which is related to a partially functioning mind. Both a partially functioning mind affirming transcendence as well as immanence, and one which affirms only immanence and denies transcendence relate to merely ideological stances based upon partially functioning mind derived conclusions. Any so-called logic which relates to either of these two stances is of a limited, partially functioning mind and not at all related to the fullness of insight and understanding which are of a meditative mind. Only such a mind can be really aware of what *is*, awareness which relates to reality, to living truth and not to a wish or desire to know. It is only self or a partially functioning mind which engages in activity which relates to any particular ideological stance, supposedly making possible the possession of that which may be called truth but which could not be living truth if there is some path supposedly related to it. What any path leads to is not living truth but merely what a partially functioning mind projects as truth. Any projection of the mind – concept, symbol, image – is not living truth, and, when the matter is reality or the living, it is the false, the illusory.

Any so-called revealed truth is seen, choicelessly and observerlessly, on a meditative mind level, to be a mere creation of a mere fragment or aspect of a partially functioning mind – reason, will, or faith. Any thought related to what is said to be understanding of any so-called truth is something related merely to what is dead – tradition, so-called revelation, scriptures, time, the relative, the experienced – and not to the living, the eternal. Any vision to which a so-called truth might relate, and any thought which is the instrument which fashions such a vision, are partially functioning mind related, and thus they have to do with the relative, the temporal, the historical, not the living, the eternal. To believe in and to seek, as suggested by theists and others, are futile activities of a partially functioning mind.

Only to a truly silent mind does living truth come. Reason is incapable of discovering living truth. Only theists who subscribe to a pre-Kantian, hence pre-critical mode of philosophizing would think that it is possible to know reality, the transcendent, the eternal by means of what Kant called theoretical reason. Indeed, not only can one not know reality by means of so-called theoretical reason, but one cannot know it by *any* part or aspect of a partially functioning mind, be it reason or some other part. The theist tries to argue or suggest that reason makes for the living of a fully human way of living, but such a way of living relates not just to a partially functioning mind, for example, one which reasons, but rather to the totally functioning mind, mind which is deeply, profoundly meditative, one which does not disallow necessary partial functioning of the mind but also does not imagine that a mind which functions partially is the totality of the mind.

A partially functioning mind, one which functions in relation to reason, to faith alone, or to faith and reason combined cannot discover reality, what *is*. Reason can arrive at partial, fragmentary, mathematical and scientific truths, that is time-related truths, but there cannot be what the theist refers to as metaphysical truths. Truth which is living relates only to a meditative mind and not to one which functions to create the illusions referred to as metaphysical truths.

One can speak, for example, about partial, objective, abstract truths of a scientific or mathematical kind. Absolute truth, related to what actually *is*, however, is living and cannot be sought out. When a partially functioning mind projects what it might call absolute truth, this is, at best, merely the fashioning of an abstraction of what actually *is*. It is futile to seek, to search for a definitive answer to the question of life's meaning, but to a truly meditative mind, one which is not functioning partially at all, the supreme, the immeasurable, the eternal, that which might be called the definitive and ultimate, can come. That which is definitive, ultimate, however, cannot be a truth recognized as something of the known, the conditioned, the relative, the temporal, something related to a partially functioning mind, and this because it is something living and not something inert or dead, hence is something new and unknown, the really mysterious. Living, absolute, or ultimate truth touches only a meditative mind, never a partially functioning one.

Those who say they know or that they can come to know absolute truth do not and cannot know it in the ordinary sense of knowing, in the way that one knows something on the level of the temporal, the conditioned, the relative. One who would look to another, be that guru, priest, or psychologist to tell one, with assumed authority and certainty perhaps, what this truth is looks in vain since what is truly significant, namely living truth, comes only to the mind when there is that loving attention which is not at all some mere activity of a partially functioning mind. Out of such attention alone comes that self-knowing which is meditative mind awareness of what the self is and what it does, that it is not anything real but rather a mere fashioning of a partially functioning mind.

Living truth is not objective nor anything subjective. The objective and the subjective are not the real, and thus any philosophy which is objective or subjective, which is realistic or idealistic is not related to what *is*, the real, but rather to what is unreal, perhaps to what seems to be, what ought to be or should be, to what is merely imagined to be.

Any synthesis fashioned or produced by thought, no matter how simple or complex, how lofty or how unadorned, is merely something manufactured by a partially functioning mind, and so it can never equate to the fullness of understanding and insight possible when there is the state of the truly meditative mind. A partially functioning mind can seek the truth, but it will never find or discover living truth. What it finds could only be mere projection out of itself, quite possibly the unreal, the illusory, the false. So-called revelation which comes to a partially functioning mind, and which is only related to such a mind, cannot be radically new. What is actually, really new is that with which a silent mind can commune, the living, that which is never continuous, never static but rather always changing and ever dynamic. So-called revelation, said by theists to relate to sacred writings, exclusively so in the case of fideism, and in union with a religious tradition which prizes reason as well as faith as in the case of theism, relates not at all to the new but rather to the old, the dead – the past, memory, experience.

1.6

John Paul II: The human being ... can find fulfillment only in choosing to enter the truth, to make a home under the shade of wisdom and dwell there. Only within this horizon of truth will people understand their freedom in its fullness and their call to know and love God as the supreme realization of their true self.

Commentator: Living truth cannot be known, cannot be grasped by a partially functioning mind, and thus a self or a partially functioning mind cannot enter into it. Such a mind cannot, by its effort, attain wisdom. When self, even if called the true, higher, or deeper self, is there to choose to live in accord with so-called truth, so-called wisdom, there cannot be full freedom of the mind. Such freedom is vouchsafed only to a truly meditative mind, and, when there is the state of the meditative mind, there is no self there at all but rather only that attention without self which is truly selfless love. When the theist talks about knowing and loving God, what he is really talking about is a mind projected something – something unreal, illusory, imagined, merely produced by a partially functioning mind – which is said to be known and loved. If a so-called higher self is not anything real, and indeed it is nothing real, then to seek to realize this so-called self is to seek to realize nothing. Such seeking cannot be anything more than irrational and futile activity on the part of a partially functioning mind.

Those who speak about the search for truth also speak about a path or paths to truth, to wisdom. There is no path, however, to living truth, to profound wisdom. Reflection, moreover, is a mere activity of a partially functioning mind, and it relates not to deep, profound, and living wisdom but rather to what is not at all related to such wisdom – experience, tradition, the past, memory – in a word, to the dead and not the living, to the stagnant and the lifeless and not the vibrant and living.

One who thinks or imagines that there is a path to living truth is one who is captive to a partially functioning mind, and the things such a mind produces, called wisdom perhaps, are not anything real or absolute but rather are unreal, false, illusory. To be enamoured of a partially functioning mind in any one or all of its forms – reason, will, imagination, and so on – and of the fashionings of such a mind is not to be one who lives in relationship to living, eternal truth but is rather to be focused on the merely relative, the measurable, even the false, the illusory. The so-called peace which is there when self ends its observations, its thinking, and its partially functioning mind related activities is not the peace which comes only to a meditative mind. Any so-called joy which this self, this so-called soul or spirit, thinks it is enjoying is not the real joy which comes to or touches only a truly silent, meditative mind.

Love and truth which relate to real wisdom are not there when self is there. Though there is often talk of a loving self seeking for the truth, real love and living truth are only there when self and all its numerous activities are negated by the attention and understanding which are of a meditative mind. Consciousness and self-consciousness are there when there is self, and real happiness is not something that is there when there is self, consciousness and self-consciousness, when there is meditation with a meditator but rather only when there is meditatorless meditation. The so-called wisdom spoken about by theists, and which is somehow related to reasoning, willing, imagining, and so on, relates to the superficial and fragmentary since it is necessarily of a partially functioning mind. It is not the deep wisdom, insight, and intelligence which can be said to characterize a truly meditative mind.

Any thought, including that related to so-called biblical, revealed wisdom, cannot express wisdom in its fullness. Wisdom in its depths and fullness relates to what actually *is* and not to what is merely thought, and only a truly meditative mind directly faces what *is*. Thus, it alone can be touched by real wisdom in its fullness.

So-called wisdom is merely something created by a partially functioning mind, whether by reason, faith alone, or faith and reason combined. It merely relates to tradition, Christian or otherwise, to the past, memory, experience, and thus it has nothing to do with the real universality of living truth. The amalgamation of faith and reason suggested by the theist is not at all related to living truth but rather relates only to a path to illusion. Any pattern effected by one or several aspects of a partially functioning mind cannot lead one to real truth, living truth, wisdom.

Theists, and others, speak of knowledge which eventually leads to wisdom. Knowledge, however, never can evolve into the profound wisdom of the truly meditative mind. Knowledge and the consciousness and partially functioning mind to which it relates must cease if there is to be wisdom beyond that so-called wisdom related to faith and/or reason. Intellect, reason, by its very nature, is incapable of exploring reality. Only a meditative mind can understand or commune with reality. A partially functioning mind cannot know the unknown, and thus what is of the known – concepts, conclusions, and so on – cannot gain entry into what the theist might call the mystery which is God but which is not living mystery, the eternally alive, the absolute, the eternal. His or her so-called mystery which is God is merely something fashioned by his or her own partially functioning mind.

Neither philosophy nor theology has ever been real wisdom, and so with regard to it there cannot be any real profundity. There is no path – philosophical, theological, or some other – which can lead to deep meaning in life. Such meaning can come only to a completely silent, meditative mind. That which alone can provide real unity and integral insight and understanding, as well as make possible a living in accord with them is the truly meditative mind. Such unity is not on the level of knowledge or so-called knowledge but is related to a level of mind beyond all knowing, to a mind which is aware now of the now and not one which projects some mere so-called ultimate goal which cannot be anything else but an unreal, illusory creation of a partially functioning mind, of a mind which functions without due respect for the limits of a properly partially functioning mind, without proper awareness of the strict limits of reason, of knowledge. Lack of such respect and awareness are there when the theist talks about speculative philosophy as that which is capable of yielding final, definitive truth. There is final, definitive truth, that is, truth which is living, but this is given or comes only to a truly silent mind and never to a partially functioning one.

Living truth is not anything objective, nor anything subjective for that matter, and the intelligence which understands living truth is not merely verbal or abstract, is not related to a partially functioning mind but rather is holistic, total, and it alone relates to the real which is not at all knowledge but with which there can be communion in the state of the totally silent mind. Such a mind is aware of ultimate, eternal values, but these are living and not abstract, and relate only to a meditative mind. They are not those so-called values which a Christian theist says are clearly revealed in the Bible, so-called values which ultimately can be nothing other than mere creations of a partially functioning mind.

1.7
John Paul II: In acting ethically, according to a free and rightly tuned will, the human person sets foot upon the path to happiness and moves toward perfection.
Commentator: Real happiness is a by-product of the loving attention of a truly meditative mind. It relates, therefore, not to consciousness but rather to awareness which is not consciousness, awareness which, because it is transcendent to awareness which relates to a partially functioning mind and to any morality and ethics which such a mind might create, does not have any dependency on an ethically informed way of living and acting. Its morality is of the deepest kind, of the kind which relates to a meditative rather than to a partially functioning mind. It is self, not anything real, which engages itself in activity related to a so-called morally perfected self. Perfection of nothing is still nothing, and so effort in

relation to perfection of self is ignorance, is futile activity, is confusion and ultimate dissatisfaction on the part of a self which is not anything real.

True, deepest freedom is not related to decision-making on the part of a partially functioning mind, of such a mind functioning morally or otherwise, not related to the exercise of a mind which merely prizes what it refers to as its freely made decisions. Freedom on the level of a partially functioning mind is a limited freedom, a freedom from something and for something else, a rejection of one thing and the positing of something else. To reject and to posit, like the making of decisions, are related to a mind functioning partially or fragmentarily, and, when such a mind which is the self engages itself in activities – choosing, affirming, refusing, and so on – there cannot be openness to reality, to what *is*, the dynamism and fullness which is life. Self-realization implies illusory movement and activity for self is not anything real, and how can one make something unreal into something real?

Faith is of a partially functioning mind, a product of such a mind's activity, and not of the action which is love, the only real action there is, and only possible when the partially functioning mind ceases its functioning and is silent. The very attention of a truly silent mind is the action of love, and, when it is there, only then is there true freedom of the mind and the living truth which can touch only a truly quiet mind and never one which is functioning partially in relationship to reason, will, faith, or imagination. The so-called truth which the theist calls real freedom is illusory, and one who lives in such so-called truth is living in the house of the dead, one built out of tradition, experience, a partially functioning mind.

True values in life, in living, are eternal rather than merely sensate. They are therefore values which one does not choose on the level of a partially functioning mind. There can be awareness of them, however, but only when the mind is in a truly silent state.

To seek to apply merely sensate values in one's life is to engage the self in futile activity which amounts to mere enhancement or aggrandizement of this same self. The truth of eternal, living values is neither subjective nor objective, hence it is not found by looking within oneself, as some people suggest, nor by seeking to grasp it objectively or impersonally, as others advise. Those who, heeding advice which is often voiced today, strive to "become themselves" or to grow as persons engage themselves in futile activity for there are no real selves to grow, evolve, fulfill themselves. "Person" is merely a partially functioning mind projection called a soul, a deeper self, a spark of the divine within, but not anything real. Living truth is requisite for full freedom of the mind, but such truth is related to a meditative mind only, not to a partially functioning one which can, and indeed often does, create illusory freedom. Total truth about man and the world comes to a truly meditative mind, not to the theistic mind which as a result of its so-called thinking concludes, for example, that it is Christian faith which makes possible a human freedom rooted in truth.[25] If person or soul is a mere creation of a functioning mind, and it is, then it is not anything real, and thus a freedom that might be said by a theist to be rooted in the human person or soul would not be real freedom. Freedom related to a partially functioning mind can only be freedom hedged in and limited, and not the total freedom that is there when there is a truly silent, meditative mind.

1.8

John Paul II: Driven by the desire to discover the ultimate truth of existence, human beings seek to acquire those universal elements of knowledge which enable them to understand themselves better and to advance in their own self-realization. These fundamental elements of knowledge spring from the wonder awakened in them by the contemplation of creation.

Commentator: Being driven by desire is characteristic of one who is ever on the level of a partially functioning mind. Self is there when there is desire, and, when the self is there, there cannot be understanding of the actual universal, that which is living, different from a merely mind-fashioned universal related to so-called elements of knowledge. Seeking by the self so that it might come to know living truth is futile activity, and any advancement in so-called self-realization is merely ego enhancement, enhancement of something unreal, a mere creation of a partially functioning mind.

Contemplation is a partially functioning mind activity. A wondering self is merely an opposite to a cynical self, and, when there are merely opposites, there is only a mind which is functioning partially, sometimes rationally but often irrationally, only a mind which is busy in relation to one of its innumerable activities. Any such activity is not action of a meditative mind, the only mind which relates to depth of understanding and fullness of living.

Person or self is something unreal, and, when it searches, it cannot find living truth. What it finds is something unreal, illusory, merely imagined, or fashioned as a so-called concept or image out of itself and merely related to its own activity.

Though the theist might define the human being as one who seeks truth, living truth cannot at all be sought out. What one can say is that a human being who is deeply aware and understands profoundly has a mind which, when it functions partially, functions correctly and appropriately, always rationally and never irrationally, but a mind which can also function totally, which can be absolutely quiet, silent, meditative when it need not function partially.

1.9
John Paul II: Faith clearly presupposes that human language is capable of expressing divine and transcendent reality in a universal way – analogically, it is true, but no less meaningfully for that.
Commentator: Analogical expression of so-called divine or sacred reality amounts to a faulty attempt to go from the known to the unknown, to use a partially functioning mind in order to grasp the eternal, the absolute, living reality. Language which properly expresses that which is transcendent but real rather than a partially functioning mind fashioning of something called the transcendent is meditative mind related. Even so, however, every such expression is never the expressed. Words of Scripture, called God's revelation by a theist, do not amount to anything other than something fashioned by a partially functioning mind, and, since this is so, they are ultimately, at best, the expression of limited, perhaps even illusory notions of what God is.

Any so-called truth to which universal validity is ascribed[26] is reason and experience related, is synthetic, that is, or else it is logical, that is, analytic. Beyond all universally valid truths related to partial functioning of the mind, beyond all logic and experience is living truth with which a truly meditative mind can commune. Universality in the case of living truth is reality rather than merely thought related, is related, this is to say, to what is dynamic and alive, the fullness which is the eternally living and not to that which is inert, abstract, dead – the past, memory, experience, the known, the recognized.

Whenever there is a call made by someone for dialogue,[27] this is actually a call for talk out of different, opposed, necessarily divisive perspectives. It relates to the striving for some measure of tolerance on the part of adherents to different opposed ideologies, each one based upon an exclusive, partially functioning mind derived conclusion which accords with, for example, some so-called dogmatic truth, some so-called deep conviction, a conclusion which is opposed to some other conclusion or to other conclusions likewise derived. After the give and take of the dialogue, divisions are still there, pasted over perhaps, and the participants in the dialogue then continue their journey down different, opposed paths. Any such dialogue can never lead to living truth but only to projections from out of partially functioning minds which are related, not to the real but rather to the unreal, the imagined, the wished for and longed for. There is no path to the truth which comes only to the mind which is completely silent, alert in its passivity, so alert that to it come ultimate understanding, insight, and wisdom, those things which alone can make for holistic living, living with a real depth of love and joy, living which eludes those who live only out of a partially functioning mind and in relationship only to what such a mind projects out of itself.

Living truth is not at all related to time and time-related cultures. It has nothing of the relative, the conditioned about it, and it cannot be known in any ordinary sense of knowing. It is ahistorical and eternal, not at all time and history related. Indeed, the mind can be aware of it only when time and history cease, are ended, but not at some future moment in time, but now, on the instant, in choiceless and observerless attention to what *is*.

Concepts are always related to a partially functioning mind, and so any universality that is theirs is not anything real but merely something thought related, rightly so when they are logical, mathematical, or scientific in kind, and when they have to do with the practicalities of living, but not when they relate to so-called religion. They have limited value when their relationship is to ethics and psychology.

1.10
John Paul II: It is rightly claimed that persons have reached adulthood when they can distinguish independently between truth and falsehood, making up their own minds about the objective reality of things. This is what has driven so many inquiries, especially in the scientific field, which in recent centuries have produced important results leading to genuine progress for all humanity.

Commentator: Science has made possible technological and practical progress for humanity. Living truth, however, relates not to so-called objective reality, something merely fashioned by a partially functioning mind but rather to the fullness of reality, reality not bifurcated as objective and subjective, to the totality of what *is*, that which can be understood and communed with only when there is that state of mind beyond a partially functioning one, beyond one which relates to duality of the observer and the observed. That state of mind is the truly meditative one. If and when, on the part of most if not all people, there is living out of the truly meditative mind, with mind functioning partially whenever necessary, rather than living exclusively, as is now the case, out of a partially functioning one, then there will be in the world not just scientific and technological progress for humanity but psychological progress as well.

True wisdom of the mind is beyond all partially functioning mind activity, activity which results in scientific and technological progress and in the religious and ethical constructs fashioned by a theistically functioning mind. Such wisdom is related to living truth and not to truth which can be sought out and found by a partially functioning mind, that is, truth which is intellectual or verbal in kind. Questions related to mystery or the unknowable unknown are answered only when there is the meditative mind which alone communes with living truth, mind which is other or deeper than a forever restless, partially functioning mind.

1.11
John Paul II: We are faced with the patent inadequacy of perspectives in which the ephemeral is affirmed as a value, and the possibility of discovering the real meaning of life is cast into doubt. This is why many people stumble through life to the very edge of the abyss without knowing where they are going.
Commentator: Discovery of meaning in life is not at all possible when there is merely the partially functioning mind. This is the case whether the partial functioning is from out of a perspective which affirms the merely temporal and transient or out of one which affirms the absolute or the eternal. That which is real, the eternally alive, the absolute, the immeasurable is not anything which can be affirmed or denied but is that with regard to which there can be communion when the mind is in a truly meditative state.

Indeed, to affirm only the temporal and the transient to the exclusion of the eternal is narrow focusing of the mind, but such focusing is likewise there when there is mere projection from out of a partially functioning mind of some imagined or so-called deeper meaning in life. Those who live in relationship to any affirmation of the merely transient and ephemeral are in no worse or better position than those who in opposition to such people choose or create a perspective which is said to be eternal, but which is really illusory, false, unreal, merely related to what might be, could be, or should be and not at all to what *is*.

Theism, a type of thinking which strives to investigate what is called being, is a form of the partially functioning mind as much as is agnosticism, relativism, or skepticism. Beyond all these forms of the partially functioning mind, indeed beyond each and every form of such a mind, is the truly meditative mind, the only state of mind which relates to fullness of understanding and holistic living.

Philosophy in any form, as it is now, and as it has been in its long history cannot directly relate to human living. Such relationship is there only when there is the state of the meditative mind, a non-dualistic state of mind wherein the observer is the observed, when there is no observer separate from the observed and not the mind which functions in relation to the duality of observer and observed. A partially functioning mind can successfully pursue something relative and time-related, but it cannot deal with any question which has to do with living truth, with what is real and not the ideal which is merely something imagined, projected out of a partially functioning mind. Only when there is awareness related to a meditative mind can there be understanding of living truth and the realization that search for meaning in life by mere partial functioning of the mind is futile, that the limits of reason and of the knowledge and so-called knowledge which it creates preclude the finding of answers to questions about the meaning of so-called personal and communal existence. Hope that philosophy in its current and past forms, all of which are/were partial and fragmentary at best, will be able to give answers to life questions is illusory. Such, however, is no reason for despair, that which is merely the opposite of hope, for awareness and understanding on the level of the meditative mind are possible, and to mind on that level answers to life, love, and relationship questions can come.

To talk, as a theist does, about God as one who would neither deceive nor desire to deceive[28] is to personify God, to personalize God, to project God as a creation of a partially functioning mind. Anything derived from such a projection of the mind, nothing at all real, will likewise be unreal, false, illusory. Thus the so-called faith about which the theist speaks, which supposedly takes one beyond an autonomous reason, reason which operates apart from any faith, has no basis in what *is*, in reality, in God. There can be discovery of what *is*, reality, God, the real, the eternally alive, of that which is immeasurable but only when there is the state of a truly meditative mind.

Any and all forms of realism and of idealism in the history of philosophy, of positivism, nihilism, pragmatism, utilitarianism, and so on are all creations of a limited, partially functioning mind. Each is a mere form of a rationally functioning mind. Such a mind, that is to say, reason in any form – whether autonomous or functioning in relationship to some faith which projects out of itself some visionary goal not based on fact, on what *is* but rather on something imagined, wished for, or longed for – cannot know living truth, and, if and when it seeks for and strives to conceptualize the absolute, it engages itself in futile activity. Also futile is a faith which is more right-brain than left-brain related, and which thus emphasizes feeling and personal experiences rather than reason and its constructs, even as it takes great delight in the images and symbols supposedly related to real insight and understanding which it projects out of itself. Futile, likewise, is fideistic faith which relates to a partially functioning mind derived conclusion which is based on biblical faith alone. Any and every partially functioning mind derived conclusion does not relate to the depth of insight and understanding possible only when there is a truly meditative mind.

Living truth is not something created out of consciousness by a partially functioning mind. It is not anything intellectual because intellect is by its very nature limited and fragmentary, hence incapable of grasping the fullness which is living truth, and because reality is not objective anymore than it is subjective. Both the object and the subject, the knower and the known, the observer and the observed are creations of a partially functioning mind. Object separate from subject or observer from observed has some point when there is knowing on a partially functioning mind level which is related to science, mathematics, and so on, but not when the concern is the living, when the matter is one of relating to others, to the absolute or the eternal.

Degrees of certainty and validity[29] can very well apply when there is appropriate or right functioning of reason in relationship to logic, science, mathematics, and practical aspects of living. These do not apply when the matter is that of living, of relating. Then what is appropriate is the truly meditative mind, mind completely transcendent to, hence beyond the limitations and the concerns for certainty and validity related to a partially functioning mind.

What some call a contemporary "crisis of meaning"[30] cannot be properly addressed by any partially functioning mind, one which engages in mere ideation, religiously based or not. Perspectives in life, whether of a scientific, theological, or some other kind are always necessarily related to the partial, the fragmentary. Any theism or fideism, as well as every type of skepticism and nihilism are all of a partially functioning mind, and so anyone of these does not and cannot lead to the discovery of deep meaning in life. Beyond the partially functioning mind, however, is the silent mind to which alone fullness of meaning in living can come. To it alone relate all of the eternal values which constitute meaning in living, in relating – eternal love, joy, freedom, and peace.

Beyond all thinking, including that which the theist says is exclusively immanentist and thus too limited, including that which the theist herself or himself prizes because it is informed by a concern for the so-called transcendent,[31] is the really transcendent, the living, the eternally alive. What the theist calls the transcendent is a mere partially functioning mind projected image or concept related not to the real, the actual, the eternally alive but only to the unreal, the inert, the abstract. Respect for the proper limits of reason is related to recognition or realization that reason is merely a mind part, a mind fragment, properly related to logic, science, mathematics, and the practicalities of living, and to the realization that it has nothing to do with truth which is living. Theistic talk of a search for truth and meaning in life is not that which relates to living truth. A theistically searching, partially functioning mind will produce what might be called truth, but this is merely the false, the illusory projected out of a partially functioning mind.

Whatever thinking is involved in the exploration of the Bible or of some other so-called collection of sacred writing, this thinking is in every instance mere activity of a partially functioning mind as much as was the thinking which informed the actual writing of so-called sacred books. Mind which relates merely to truth or so-called truth of a merely intellectual kind and to statements which express such truth or so-called truth is a partially functioning and not a meditative mind. When the mind functions partially, its functioning is proper if the matter at hand is science or technology, for example, but not when the functioning is projection of what the mind is merely imagining the eternal, the immeasurable, the ineffable to be, and projection of constructs intended to encase what it really is.

A philosopher of being engages his partially functioning mind in a search for what he calls objective being. Reality, however, is not something objective, and, when there is so-called philosophy of being about which the theist speaks a partially functioning mind is engaging itself in activity related to fashioning of the false, the illusory, the unreal. Any philosophy of the past or the present was/is partially functioning mind related and so could not/cannot attain to something absolute, ultimate, fundamental. The search of a philosopher of being for truth is illusory. That which is truly good is not anything which can be known, but that of which there can be awareness only in the state of the meditative mind. Reality and truth do not oppose the factual, but rather are related to what *is*, the actual. When, in its seeking truly and certainly to know what is termed metaphysical reality, a theistically functioning mind goes beyond the factual, the so-called empirical, it is engaging itself in illusory activity for reality – goodness, truth, and beauty – does not come to a partially functioning mind, mind which sets up false dualisms and opposition, for example, subjective interiority and so-called personal spirituality on the one hand, and so-called objective reality on the other. Both a philosophy which merely rejects metaphysics and one which prizes it cannot commune with the absolute, the eternal, that which is supposedly, but not really understood by a faith based upon so-called revelation, Christian or otherwise.

Scriptures, the so-called "Word of God," written at a time, perhaps, when right-brain rather than left was developing relate to a partially functioning mind, and so any theistic call for transcendence beyond an autonomously functioning reason amounts to mere projection out of a partially functioning mind. Such projection never relates to what is real, living, eternal but rather to what is unreal, the merely imagined, what is wished for, desired. Though human *a posteriori* knowledge is properly limited to sense experience, there can be, on the level of the meditative mind, communion with mystery, the unknown which cannot be known, the eternal, the absolute. Metaphysics has been and is a futile attempt of a partially functioning mind to grasp what truly transcends such a level of the mind. Theology, too, whether related to metaphysics or not, is related to partial mind functioning, and so it cannot speak meaningfully of what is beyond the temporal, the experienced, that is, the unknown.

A merely convinced partially functioning mind functions wrongly when it follows the path of metaphysics for there is no path at all, this or some other, to living truth, communion with which is the only real and significant movement beyond the shallowness and superficiality of contemporary ways of living. Something which needs to be said is that it is a lack of meditative understanding, insight, and wisdom which is the cause of any crisis of meaning in this our contemporary world.

Theists, like John Paul II, speak about the value of hermeneutics and language analysis as regards the clarification of structures of thought and language,[32] but they bemoan the lack of effort on the part of philosophers who prize hermeneutics and language analysis to search out and find the so-called essence of reality. What these theists are actually trying to do is breathe new life into a traditional, pre-critical type of philosophy, a so-called philosophy of being, hoping that it will be able to do what philosophers starting with Kant have often shown to be a futile and impossible task, namely, to know reality. What both philosophers of being and post-critical philosophers could do is talk about mind stopping its partial functioning at some point, at that point where partial mind functioning is no longer appropriate, so that reality, living truth, that which is really the transcendent might touch the truly meditative mind. All thought and activity related to hermeneutics and analysis of language, as well as every partially functioning mind effort to go beyond them must cease if there is to be true communion of the mind with reality, the unknown, the eternal, the immeasurable. Reality has no merely mind projected essence to be known by a partially functioning mind, but of it there can be awareness which is not consciousness, that is, awareness without a knower seeking to know the unknown, without a subject facing an object. Such awareness is communion with reality, the unknown, that is, what cannot be known. Such awareness is understanding on the part of mind in a truly meditative state.

Historicism, scientism, and pragmatism[33] all relate to conclusions arrived at by partial functioning of the mind. Beyond a partially functioning mind, and hence beyond any and every conclusion or stance related to it is the truly meditative mind. Any formulation of a partially functioning mind relates not to living truth but rather rightly or wrongly to what is truth, or what is called truth, of a merely intellectual kind, of a kind which always has to do with the relative and not the absolute, with time and not the eternal. Metaphysical statements are meaningless, but metaphysical questions are not for they are susceptible to answers on the level of a truly meditative mind. It is the actual, what *is*, that such a mind sees, without self there seeing, choosing, hence distorting the observation of what *is*. When there is such a state of mind, science has its proper place but is not corrupted into scientism. Then neither, however, is there metaphysical speculation related to partially functioning mind projections which some, likely a minority today, suppose can make for a proper address to the truly significant questions in life. To equate the human being to the merely rationally functioning mind as all ideological thinkers and all philosophers of being do is to narrow down the whole of the mind to its partially functioning mind parts, one of which is reason or intellect.

A theist's judgments founded on religiously ethical principles[34] relate to a partially functioning mind, hence to the merely relative and temporal, and not to the really eternal values – love, freedom, peace, order – of which a truly quiet mind is aware. Any theistic vision of what the human being can become relates to the unreal, the merely imagined, to something merely mind-projected. Rather than vision which includes or excludes solutions to the moral dilemmas of the contemporary world and the overcoming of the evils and sufferings of this world, and rather than the imagination and speculation which relate to the construction of any vision, there can be the loving attention of a truly silent mind and the loving, total understanding and intelligence which come from that attention, and as well a way of living in accord with such understanding and intelligence. Only the latter, and not the former relates to the real, the what *is* rather than to the unreal, the merely imagined, what is merely wished for and longed for.

Living truth and full freedom of the totally unconditioned mind do not at all relate to a philosophy of being and its so-called objective truth, nor do they relate to the atheism, nihilism, and agnosticism which are opposed by a philosopher of being. The truly silent mind is mind alone, is therefore neither theistic nor atheistic, neither deistic nor nihilistic, neither fideistic nor postmodernistic. Such a mind neither hopes nor despairs, nor does it constantly walk in the hallway of opposites. Even though unattached, and not identified with any ideology or system of belief, however, such a mind is not a lonely mind for it communes, moment to moment, day to day, with the eternal, the immeasurable, the unknown, living mystery.

Only when there is never the state of the truly meditative mind is there living without deep and ultimate meaning. Such meaning, however, is not the so-called deeper meaning about which a philosopher of being speaks. When there is the deep meaning which is of a truly meditative mind, one is beyond any philosophy of being which the theist says is the necessary counter to modes of thinking which relate to what he says is an absence of meaning in the contemporary world. When in a truly meditative state, mind is beyond any and every ideological stance or perspective, hence beyond the partially functioning mind which creates any and every such stance or perspective.

Such, then, is one example of Western left-brain religious thinking. Such thinking, to put it briefly, would respond to the first of the two questions posed in the introduction to this book by saying that yes, indeed, a rationally functioning mind, provided it works in union with faith, can lead to the possession of a truly religious mind, or rather such a functioning mind is the truly religious mind. Enough indications have been given in the commentary on passages from *Fides et Ratio* to show the inadequacy, indeed the falseness of such an answer. What was perhaps the main thing said in this work by John Paul II as regards an answer to the second question was that, though science has its importance, something more than science is needed to adequately address what has been called the crisis of meaning in the contemporary world. This something, of course, is faith and its partially functioning mind creations. Such an answer has already been said here to be inadequate.

In the second half of this chapter, we see an attempt on the part of Anthony De Mello to address what John Paul II calls the contemporary crisis of meaning by his presentation of what might be said to be predominantly right-brain thinking on the level of the partially functioning mind. By staying always on this level of the mind, it is being suggested in this book, the two questions already posed cannot be adequately answered. More about this later, however. Let us give our attention now to some of the things De Mello says.

1.12
De Mello: It is not from lack of religion in the ordinary sense of the word that the world is suffering, it is from lack of love, lack of awareness .... Turn on the light of awareness and the darkness will disappear.
Commentator: There is no way to awareness on the level of the meditative mind, no method, no system which will lead to the light. When there is a way, or a path, self is there, or selves are there engaging in partially functioning mind activity. Only when mind completely ceases its partial functioning, its activities, can the light, that is, the unknown, the immeasurable, the absolute, the eternal come to the mind. All ways, all methods must cease, and then awareness will come, if it will, so to speak.

What De Mello calls awareness here is awareness which is consciousness. It is not observerless awareness. Self is there controlling, thinking, cogitating, generating love, hoping, wishing, wanting, and so on, and, whenever self is there, there cannot be total understanding and total freedom of mind alone, mind completely uninfluenced, mind utterly and completely unconditioned. Love, moreover, is not anything generated, fashioned, or created by self but comes when there is emptiness, complete silence of the mind, hence no activity to get, or to acquire it.

A fundamental difference is that between a partially functioning and a truly meditative mind. What one by and large sees in the world today, however, is exclusive engagement in partial functioning of the mind. People today do distinguish between discursively left-brain and what might be called poetically right-brain thinking. People like John Paul II show a preference for the former, at least as far as religion is concerned, and those like De Mello demonstrate a preference for the latter, but there is seemingly no indication of the profound importance of meditative mind awareness, awareness which relates to a level of the mind beyond that on which there is left-brain and/or right-brain thinking, awareness which is deeper than that which relates to any partial functioning of the mind. Such awareness is necessary if there is to be a truly religious mind while awareness which entails presence of the self or of a partially functioning mind is superficial, no matter how necessary at times, and how unnecessary at other times, it is.

1.13
De Mello: The concept is a help, to *lead* you to reality, but when you get there, you have got to intuit or experience it directly .... To know reality you have to *know beyond knowing.*
Commentator: To say that a concept can help one to reach reality is to suggest that the known can reach the unknown, the measurable can encapsulate the immeasurable, that there is a way or a path leading to reality. Concepts must cease or be completely negated if reality is to come to or to touch the mind.

Where one "gets," with concept as a vehicle to get there, cannot be reality but rather only something unreal, something created by partial mind functioning, something merely imagined perhaps. When there is intuition out of a truly meditative mind, there is only experiencing, not a self intuiting, experiencing. Concepts and ideas cannot lead to reality. They must be ended, negated, if reality is to be.

Really to be aware beyond knowing, beyond the consciousness to which knowing is related, and beyond all of the words and concepts to which such consciousness and knowing relate, is not to *know beyond knowing* in the sense implied by De Mello. Beyond knowing is self-knowledge, but the latter requires a truly silent mind, that mind which is never talked about by De Mello.

Strictly speaking, there is no such thing as knowing beyond knowing. To say or imply that there is shows a lack of understanding as regards the limitations of knowledge, of the reason or intellect which fashions knowledge. There are different forms of knowing, perhaps, different types of knowledge, but all are basically or fundamentally on the same level, the level of a partially functioning mind. For the deep awareness which is not consciousness, which is awareness without self there knowing something or anything, and be that self the poet, the mystic, a so-called great philosopher, or some other, the knower and the known must cease, be ended, be negated by choiceless, observerless attention to what *is*. Then only is there reality which can touch mind in a truly meditative state, mind that is absolutely silent, totally open and receptive therefore. To help one arrive at reality De Mello outlines what he calls steps to wisdom.[35] There cannot, however, be steps to real wisdom. There is no path of any kind to living truth.

One of these so-called steps to wisdom relates to the suggestion that a person – what De Mello would say is one's real self – should never identify with any negative feeling.[36] It is illusion, he suggests, to define oneself, that is, one's so-called true self, with any negative feeling. The fundamental illusion at work in what De Mello says on this point is to think that there is a real, so-called essential self which either does or does not identify with, and which either does or does not define itself in terms of some negative feeling. What must be realized in relationship to any such feeling is that the observer is the observed, the watcher is the watched. Thus, the one who feels loneliness or envy is the loneliness or the envy, not a something real which can either identify itself with a feeling or not. Not to realize this is to be in illusion, not to be wise with the depth of wisdom which is given to a truly meditative mind, wisdom which relates to the deep understanding that self in any form – either higher or lower – is not real, and, since this is so, there is no point in identifying or not identifying what is unreal with something one feels, nor at all in attempting to define what is unreal.

1.14
De Mello: Another illusion: You *are* ... labels ... put on you ... what you need to do is smash the label! Smash it, and you are free!
Commentator: A labeless self is still self as much as is a labeled one. If there is smashing of a label, this is an activity of a partially functioning mind, of self. Real freedom of the mind never comes as a result of any such activity but only when there is no activity of the partially functioning mind at all and no self at all to engage itself in any activity. Self living "life from moment to moment," as De Mello suggests, is not selfless living from moment to moment.[37] Smashing labels is one example of self paring off the layers of accumulations until there is only the so-called real self, the "I," the soul. "I," the soul, the so-called real self, however, is nothing real. Real understanding of what the "I" actually is comes only from observerless observation of this "I" which engages itself in its numerous activities. Out of such observation comes self-knowledge, which is not a knowing by a partially functioning mind of what is believed to be a real self. Profound wisdom is the realization that there is no real "I" or "You"

to be touched by things in life which can upset one, which can cause loneliness, depression, and so on. There is only loneliness, depression, despair. When there is looking at something – loneliness, sorrow, and so on – without the looker there looking, when, this is to say, the observer is the observed, the thinker is the thought, the watcher is the watched, then the mind is totally aware and fully understands, and it is this totality of awareness and fullness of understanding which are the very dissipation of the observed loneliness, the envy, and so on.

Not to identify a so-called deeper self with negative feelings is an example of what De Mello would call enlightenment.[38] Any such enlightenment, however, is not true enlightenment for here there is no awakening of the intelligence and understanding which are of the meditative mind. Such enlightenment brings the realization that there is no real "I," no soul, either to identify or not to identify itself with something. What De Mello would say is a radical difference between the self which identifies with negative feelings and one which does not[39] is merely the difference between two opposites.

The peace which supposedly comes to the self which no longer identifies itself with negative feelings is illusory, is not that peace which comes to or touches a really quiet mind. Such peace is merely a mind-projected peace and so not at all related to the real, the true. Moreover, it is a partially functioning mind which would advise, urge, and thus, if and when mystics urge others to engage in partially functioning mind activity so as to discover a so-called real "I,"[40] theirs is not at all the deep understanding and insight that are of a truly meditative mind. Indeed, what they are really doing is leading others down a path to illusion, a path down which they undoubtedly themselves have walked or are walking.

There is profound wisdom and an enlightened mind, not when a supposed real self either does or does not identify itself with something, be it church, nation, ambition, loneliness, or something else, but when there is the realization that there is no real, higher or deeper self that might be said by a partially functioning mind to be rightly or wrongly identified with something. Conflict and confusion are there when there is an identification of self, whether it be with negative feelings or so-called base things, or with something so-called positive or noble, like church, family, culture, and so on. To cling to what is called God, to church, to a painting, a house, or furniture is fundamentally no different from clinging to possessions, money, ambition, and so on. Identification with or clinging to things so-called spiritual is no different from identification with or clinging to so-called material things. In both cases the mind is not free but rather attached.

There is a profound difference between non-identification out of a totally silent mind and that which comes out of a partially functioning mind. The former means there is no self there identifying or not identifying. When there is non-identification by a partially functioning mind, self is there, and this self merely stands opposite to the self which does identify itself with something. Only when there is no identification by a truly meditative mind and no self, that is, a partially functioning mind attached to or detached from something, is there real love. Self loving people in general or someone in particular is still self, and the love that is there when self is there is not real love but merely what is ordinarily called love.

The suggestion which De Mello so often makes is that happiness will result once a so-called deeper self no longer identifies with negative feelings.[41] Happiness, however, is not merely an opposite to depression, upset, loneliness, striving for success, and so on. Being truly free is happiness, but true freedom has nothing to do with self being someone or something opposite to the successful person, the woman or man of importance, of high or low position, and so on, anymore than it has to do with someone being one of these forms of the self. If self is there no longer concerned anymore about money, success, and so on, then this is the mere opposite of self being concerned about them. Self not belonging anymore to a certain group is merely the opposite of self belonging to such a group. Self being a nobody is merely the opposite of self being a somebody. Self wanting not to be somebody, not to belong to anybody is merely the opposite of self wanting to be a somebody, wanting to belong to someone.

One who does not identify self with negative feelings is supposedly one who knows one's deeper self. The suggestion is that only when there is such knowledge, can there be right action.[42] Self can never really act, however. What it always does is merely engage itself in activity. The only real action is love, and there is no real love when there is self in any form, so-called higher or deeper, or lower.

What is being prized so often by De Mello is subjectivity versus objectivity, that is, the individual versus the collectivity. When opposites are there, however, self is there, and there cannot be profundity of understanding, hence of living, when self is there.

Any so-called deeper, subjective self is as unreal as the so-called realities which the self creates, that is, projects out of itself. There is no actual, real self at all to be found by a so-called losing of a so-called ego or lower self. Any so-called realization that there is a higher or deeper self which is not the same thing as the lower or superficial self is merely projection by a partially functioning mind. A soul, a god-within, a higher or deeper self, a so-called non-centred self is still self. Deepest realization relates to the understanding that one is "nothing," that is, not a thing like a so-called soul or real self, to the understanding that there is no soul or real self at all.

Person is self, and so-called self-worth or personal-worth is mere projection from out of a partially functioning mind, not related at all to anything real. Identification of oneself – supposedly, but not really something real – with things which are said to be temporary or ephemeral is fundamentally no different from non-identification with temporary or ephemeral things. When there is the state of the meditative mind, there is real communion with the eternal, the living, and no identification at all, that is, no self either to identify itself with something or not.

Person is not anything real, and so to experience, to so-called "intuit"[43] person is to experience what is not real, something which is merely mind-projected. Person is not beyond the thinking mind but is there as a creation of or something made up by such a mind. What is beyond both person and the thinking mind which creates person is the meditative mind and the absolute, the immeasurable with which such a mind communes.

Concepts are important in science, mathematics, technology, and as regards the practical aspects of living. The intelligence to which they relate, however, is merely verbal or theoretical. It is not the intelligence of a meditative mind, total and profound as this is.

De Mello suggests that one resort to myth, to the images and symbols associated with myths, in order to express so-called insights which are said not to be possible so long as there is merely conceptual thinking.[44] To do this, however, is to engage a merely imaginatively functioning mind, is to go from one stage of consciousness development to another, from the fragment of the mind which relates to this one stage, namely intellect or reason, to another mind fragment, namely, imagination which, as regards its initial, historical development, very likely came before the full flowering of reason.

The structured wonder of the mystics[45] – structured, that is, by a partially functioning mind – and the so-called intelligence which relates to it have to do, not with reality, not with what *is*, but rather with what is merely imagined and so unreal. Any effort to acquire again a sense of wonder which was supposedly there at an earlier point in one's life but then subsequently lost, for whatever reason, is to engage the mind in illusory activity. Related to what might be said to be a way of understanding which is more mystical than conceptual is the process of abandoning concepts and judgments,[46] things related to a left-brain partially functioning mind. Implied, however, is not mind stopping all of its partial functioning so as to be really quiet, but rather mind functioning partially, that is, self abandoning concepts and judgments. Once the abandonment has occurred, self is still there, and, when self is there, there cannot be the clarity of perception and fullness of understanding which are of a truly meditative mind.

Our minds are very much conditioned in many different ways. Some of these ways are outlined by De Mello.[47] The sensate values related to such conditioning, however, do not lead, as he suggests, to a loss of a real soul for soul is not anything real, hence not something which could be lost. What exclusive prizing of such values means is that lives are lived very superficially, totally related to a partially functioning mind, mind functioning rationally and/or irrationally, without any awareness that the eternal values are the ones which alone relate to a profoundly meaningful way of living.

Self effecting unconditioning of the mind will not make one's living deeply meaningful for once the unconditioning is done, self is still there, and self will sooner or later bring into play other, perhaps more subtle forms of conditioning, will bring conflict, confusion, even chaos into one's way of living. If self is there not making an effort of will, if self is there without any ideals, plans, or schemes, if self is there without habits, self is still there, and, when there is self, there cannot be awareness which is not consciousness and the deep insight, understanding, and wisdom which constitute a profoundly meaningful way of living.

When there is life truly and deeply lived, there is no striving for things called spiritual, like love, happiness, holiness, peace, freedom, and humility, all of which are not real if self is there loving and being happy, holy, peaceful, free, and humble. Spiritual striving is no different from striving for material things for, in the case of both types of striving, self is there striving. When there is deep living, then there is no prizing on the part of a partially functioning mind of the individual as against the collectivity, or vice versa, of the subject as against the object, or vice versa. Rather, there is only the individual mind in the sense of mind alone, indivisible, not on a level of opposites, not separated into fragments or parts. Such a mind is not opposed to conceptual thinking, to analysis, and so on, but it does see the proper place and the proper limits of them, and thus it is on a level transcendent to the partially functioning mind which engages itself in conceptual thinking, in analysis, and in other activities like these. It is mind beyond any and every fragment of the mind – imagination, will, reflection, reason, and so on. Only when there is such an individual mind is there no conflict, no violence but rather mind which is whole, truly meditative, truly religious.

Contrary to what De Mello would say,[48] a truly empty, soulless way of living is profoundly wise living. Such emptiness, however, is not opposite to a fullness made up of so-called good ideas and values which are opposed to so-called bad ideas and values. The transformation of mind about which De Mello speaks[49] is neither radical nor profound, hence not at all revolutionary, for it relates to mere movement from one point on the level of the partially functioning mind to another, and not to the awakening of the meditative mind's loving intelligence and total understanding.

1.15
De Mello: Time ... given to worship and singing praise and singing songs could ... fruitfully be employed in self-understanding.
Commentator: Living understanding of the self is understanding of what the self really is, that is, nothing real. Such understanding is not at all time-related as is so-called verbal, intellectual understanding. Any effort to acquire so-called self-understanding, so-called understanding of a so-called real, but really unreal self, is as fruitless as worshiping and hymn-singing. Such effort cannot ever result in self-knowledge, that is, knowing on the level of a meditative mind of what the self is and what it does, sometimes appropriately but often not, and that it is not anything real.

When there is self, there can be dealing in terms of opposites. Self-satisfaction, for example, is merely the opposite of self-dissatisfaction, and, whenever self is there, psychologically speaking, there is a problem. Self deciding not to make effort, to be contented, not to exert itself in effort to attain is engaging itself in activity as destructive as that related to self which is dissatisfied, discontented, and always striving for what it foolishly thinks will bring contentment and peace to itself. In either case there is conflict and so no quiet mind understanding. To understand all of this is to understand the self, to know the self, moment to moment, day to day.

1.16
De Mello: If you wish to love, you must learn to see again.

Western Religion, Science, and the Meditative Mind 45

Commentator: Is this seeing again – seeing in a so-called new and fresh way – a seeing with the same old eyes? Is the seeing merely related to a mind which has moved from one type of partial functioning to another, perhaps from left-brain to right-brain functioning? These are the questions which have to be asked here, and if the answer to both questions is "yes," then the seeing is not the total seeing which relates to the truly meditative mind.

Here, the "seeing again" seems to imply seeing in a way that is different from the way most other people see. The implication seems to be that self must withdraw, be alone in order to see again. Self alone, withdrawn, however, is never really alone, that is, its aloneness is not at all the aloneness of the truly meditative mind. When there is aloneness of the meditative mind, there is mind totally unconditioned, that is, totally free. When there is this real aloneness, self is not there, and so love is there.

There is seeing or observation with observer, self, or see-er there, and there is seeing without the observer, self, see-er present. The latter kind of seeing is choiceless, observerless seeing. The former type of seeing is De Mello's seeing which does not equate to totally unconditioned seeing. Self which sees means some form of conditioning in relationship to past experiences is there. A truly alertly seeing mind is not disciplined in the sense that there is exercise of control or concentration of the mind. When the partially functioning mind sees, it sees only what is dead and not what is new and fresh, not what is of the unknown but merely what is of the known.

What De Mello talks about is observation with observer there,[50] and not about observerless observation. Always self is there observing. Always there is the dualistic structure of observer facing the observed. He never talks about seeing out of a meditative mind, that seeing which is related to awareness which is not consciousness. In De Mello there is only awareness which is consciousness, awareness which implies presence of the self.

To break free of a prison of concepts, judgments, opinions, and so on, without movement beyond a partially functioning mind is to be in some other prison, perhaps the prison of the mystic whose partially functioning mind creates images and symbols rather than concepts, whose language is right-brain rather than left-brain related, or the prison of the soul, that is, the unreal self which engages in spiritual exercises and disciplines.

Total clarity of perception is impossible as regards perception by a partially functioning mind.[51] Any perception by such a mind involves dualism of the perceiver and the perceived. Perceiver is self, and, when there is self, there cannot be that total clarity of perception which is possible for a truly meditative mind, mind without self there to skew or distort perception, to a greater or lesser degree, but to some degree nevertheless. Simply to desire clarity of perception, or to think that there is value in aspiring to reach it, is indication that one is ever and always on a partially functioning mind level.

When there is nondualistic mind perception, and this out of a meditative state of awareness, the understanding which comes from such perception is the only response needed to any question related to living or relating. Such a response, moreover, is completely right, hence absolutely accurate.

Though one can go out to get water in a bucket or sand in a bag, one cannot go out to get, acquire, achieve awareness,[52] at least if this awareness is related to seeing out of a meditative mind. Any such effort to get something like quiet mind awareness, insight, or wisdom is futile activity on the part of self in some form, is merely movement of the self as it seeks to acquire or to achieve what it itself projects as a desirable goal to be reached. What it is striving to reach or obtain is necessarily something which is unreal, something illusory. Total awareness which is not conscious, self-related awareness comes to the mind but only when it is in a meditative state, when it is seeing choicelessly and observerlessly. Seeing by an observer, thinker, or self is biased or prejudiced and prevents living truth and fullness of understanding from coming to the mind.

When De Mello advises people to see, to watch loneliness, pain, disappointment, and so on, this seeing or watching is dualistic watching. It is watching with observer there observing the observed and not observerless, choiceless observing or watching, the only kind which is not touched by any of that conditioning which militates against mind being really loving, really free.

Real love is not pleasure, is not pleasant feelings or experiences. "Pleasant experiences"[53] relate to pleasure, and to relate life to pleasures and pains, to seek in life merely the cultivation of pleasure and the avoidance of pain is to live superficially.

Many people, including De Mello, suggest that "painful experiences" can "lead to growth,"[54] but this is like saying, wrongfully, that one can go from the known to the unknown, from the measurable to the immeasurable. Psychological pain, that is, pain which is caused by activity on the part of a partially functioning mind functioning inappropriately must cease, be ended by insight out of a truly meditative mind, by choiceless observation of it, and only then is there that timeless state of total awareness and understanding which alone can be the dissipation of psychological pain. Such a state of mind is not something which one grows into for growing into something implies a time-related process, and time cannot give way to timelessness. If timelessness is to come, time and all that is associated with it must cease.

1.17
De Mello: Living is to have dropped all the impediments and to live in the present moment with freshness, [and] eternal life is *now*, in the timeless *now*.
Commentator: Self dropping its perceived impediments is related to effort of a partially functioning mind striving to attain to the state of the truly silent mind. Activities in relation to what are considered to be things spiritual and in which there is engagement in order to discover meaning in living, in relating, are as futile in this regard as are so-called good works said to help others, to build up the Kingdom of God, and so on. When there is no state of the truly meditative mind, there is no living in the newness of the present moment, that is in the eternal now, but only living in relation to what is imagined by self to be such newness.

Indeed, eternal life is in the timeless, the eternally now moment, but the crucial question is how does one approach that which is living, directly when there is mind in a meditative state or indirectly with a partially functioning mind, a mind which readily fashions images and concepts of living which are not themselves living things, but rather abstract, inert, hence, nonliving things. Whether one lives in a house of the dead or open to the new, the unknown, the living, in relationship to what is unreal and merely imagined or to what *is*, to reality, is dependent on which of these two approaches is the one taken.

1.18
De Mello: Ideals do a lot of damage ... you are focusing on what should be instead of focusing on what is.
Commentator: When De Mello talks about the harm which ideas can cause, what he says sounds very much like vintage Krishnamurti. However, what De Mello says is really quite different from what Krishnamurti says for De Mello was always on the level of a partially functioning mind when he wrote the contents of his books whereas Krishnamurti was always on the level of a truly meditative mind whenever the matter or concern at hand was one of living rather than one related to the theoretical or the abstract, as in consideration of scientific and technological problems. Self is there when there is mind functioning

partially but not when it is truly meditative. Whether self focuses on what should be or what is, there cannot be the understanding of reality that is only possible when self is not there, and it is not there when there is a truly silent, meditative mind.

1.19
***De Mello*:** When you become love ... you have dropped your illusions and attachments.
Commentator: Self can never become really loving. There is no such thing as a really loving self. Becoming implies self. Self is there before the becoming occurs, during the process of becoming, and once the goal of the becoming process is reached. Where and when self is there, there cannot be love. Self attaches itself, and self detaches itself. Attachment and detachment are activity of self, of a partially functioning mind.

1.20
De Mello: Stay ... for a while [in] the vast desert of solitude, [and] you will understand what freedom is, what love is, what happiness is, what reality is, what truth is, what God is. You will see, you will know beyond concepts and conditioning, addictions and attachments.
Commentator: So long as self is there, either isolated or not, in the city or in the desert, in the office or in the monastery, there cannot be real, meditative mind understanding of what freedom, love, happiness, reality, truth, or God is. What self sees beyond concepts and conditioning, and beyond addictions and attachments, is no more real than are these contents of a partially functioning mind.

Self without concepts is still self, and, when self is there, pristine and total seeing, seeing out of a meditative mind, is an impossibility. Where self is, there is no direct contact with reality at all. Self can move beyond concepts so that there is right-brain related consciousness and right-brain language expression rather than left-brain consciousness and its associated language, so that there is thinking in relation to images and symbols rather than to concepts, but any such movement cannot overcome hurt, pain, suffering, conflict, loneliness, and sorrow, anymore than can conceptual thinking. Indeed, it is self, with or without concepts, whether higher or lower, which is the cause or the creator of these and other problems in living.

Needing, using, manipulating, finding "ways and means"[55] are all activities of a partially functioning mind, and, when a partially functioning mind, that is "I" or self is there, there is no love, no happiness. Then there is no profundity of living.

Living without any psychological pain and conflict, without hurting and being hurt, without forgiving and being forgiven, without feelings of success, ambition, greed, or envy is possible only when there is emptying of the mind, but such emptying is not "I" or self emptying itself. This emptying is that which relates to choiceless, observerless observation of what *is* and the understanding that comes of such observation. When there is such emptying, there is no sorrow, no feeling confused, no feeling lonely for then there is no self there to be sorrowful, to feel confused, or to feel lonely.[56] Self alone or in solitude is not aloneness of the truly meditative mind, of mind which is meditative but without meditator there. Only when there is this aloneness can there be awareness and understanding of what love and reality really are.[57] Love, God, reality cannot be known, are not at all related to thinking in the ordinary sense; indeed, if these are said to be known, what is known is not real love, not God, not reality but merely mind projections of what these might be.

When thinking is there, when one gives place of primacy to thinking over living[58] rather than to living over thinking, "I" is there, self is there, that which will necessarily interfere with living, relating. Self there means there cannot be full clarity of seeing, there cannot be the attention of the truly loving mind, there cannot be real love therefore.

1.21
De Mello: Suffering is given ... that you might open your eyes to the truth, [and] insight is a great help, not analysis.
Commentator: One's eyes do not really open on the level of experience, as a result of one's having suffered. There is no goodness in suffering. Goodness is there when the suffering caused or created by a partially functioning mind functioning improperly is ended, and this by a meditative mind understanding of it. This very understanding is the ending of it.

One is in illusion when one thinks there is a real "I" different from "me," the ego. There is no psychological suffering when there is no self there, when one understands out of observerless, selfless observation, when there is quiet mind understanding that comes out of such observation.

It is self which can get psychologically sick,[59] that can look to experience as a way of overcoming suffering which is experience related. To use suffering to end suffering is like using the known to try to acquire the unknown, to use the recognized, the relative, to attain the unrecognized, that is, the new, the eternal, the absolute. If people "go on suffering,"[60] day after day, year after year, this is because there is no quiet mind attention or observation, and the understanding which comes of such attention, such observation.

Self which seeks to end suffering is the same self which suffers. Only when one truly sees the self for what it is and what it does, can there be a real end to suffering for only then is there an end to the self which can cause suffering or which can seek to escape from it. Real insight ends suffering, but meditative mind insight is not at all experience related.

1.22
De Mello: Love the thought of death.... Think of the loveliness of that corpse, of that skeleton, of those bones crumbling till there is only a handful of dust. From there on, what a relief.
Commentator: If there is a feeling of relief, self is there feeling relieved. Only when self is not there is there real love and no conflict at all.

However necessary thought at times is, to have so-called love for thought is to have it in relationship to what is superficial and relative rather than to that which is deep and absolute. Rather than loving something unreal, one can live and die moment to moment and day to day. Only such living-and-dying is fully human and significant living.

Living-and-dying everyday[61] is related to meditative mind insight and understanding. Such meditation, however, is not meditation with meditator there, a meditator who De Mello says does well when daily he or she engages in meditation which involves imagination of his or her own death.[62]

De Mello suggests that one should see death as something wonderful rather than as something to be feared.[63] To suggest, however, that one way of viewing death is better than another is indication that one is merely on the level of mind where opposites prevail. If death is either wonderful or something to be feared, self is there thinking it to be wonderful or to be feared. Total understanding of life is only on the level of a meditative mind, not on the level of mind where opposites reside, where there is self functioning in some way or other. People are not fully alive when their minds always and only function partially but when there is, significantly, the truly quiet state of the mind, which mind allows for proper partial functioning of the mind but which also realizes that besides partial functioning of the mind there is the truly meditative mind, that is, besides the superficial aspects of the mind there is the deep, meditative level of the mind.

1.23
De Mello: You do not need to belong to anybody or anything or any group. You do not even need to be in love.... What you need is to be free. What you need is to love.
Commentator: When self is there needing or not needing something, one is traveling on the highway of opposites. One's thinking is dualistic when there is a choosing of one side of an opposition, for example, society or the group as against the individual or the individual as against society, against the group. One is fully human, not when one is an individual in the sense of opposed to the group or society but when one's mind is indivisible, whole, capable of being in a state of complete silence beyond all opposites. Only such is a state of mind related to true love, true freedom, both of which are possible when there is no self at all, no "I" supposedly different from ego.

1.24
De Mello: Life has no meaning; it cannot have meaning because meaning is a formula; meaning is something that makes sense to the mind .... Meaning is only found when you go beyond meaning. Life only makes sense when you perceive it as mystery and it makes no sense to the conceptualizing mind.
Commentator: Life is filled with meaning, but such meaning is not anything a partially functioning mind might strive to fashion. Communion with what is living is this fullness of meaning. That which a partially functioning mind or self perceives as mystery is only that which itself has projected as mystery. Life perceived by a perceiver facing the perceived relates not at all to real, actual mystery but only to mystery as a projection out of a partially functioning mind. Real mystery, the living, the unknown can be perceived only when there is perception without perceiver, the watched without the watcher, when the perception or the watching is related to a truly meditative mind.

There is meaning of an intellectual, verbal kind on the level of a partially functioning mind functioning properly. Deepest meaning in life, in relationship to living, however, is not something intellectual, verbal, or conceptual. It is not something opposed to the intellectual, the verbal, or the conceptual but rather is on a level of mind totally beyond or truly transcendent to the level of partial mind functioning, the level whereon opposites reside. This deepest meaning is not anything found by self, neither the so-called lower self or ego nor the so-called higher self or soul, but rather there is discovery of it by mind in a truly meditative state only when there is no self at all. Such meaning comes to this state of the mind and is not a so-called meaning effected by some activity on the part of a partially functioning mind. The meaning beyond meaning spoken about here is not this deepest meaning but rather partially functioning mind projected things other than, if not opposed to, the concepts which that same functioning mind produces.

1.25
De Mello: Love is sensitivity, love is consciousness.
Commentator: Love is not consciousness, anymore than happiness is. When one is conscious that one is happy, happiness has already gone out of the window. So too with love.

Love is not consciousness for consciousness implies self, and, where there is self, there cannot be love. There is love only when "you" are not, when self is not there. Self must be negated, and activities of the self must cease, if love is to be. Real love is the attention of a meditative mind, is meditative mind awareness which is not conscious awareness, that is, is not self being aware of something, not even of itself.

Love has no opposite. When "I" in any form – even if it be soul, a so-called higher self – is there, there is no real love, no truly religious mind. When there is a really religious mind, self is not there at all.

When De Mello says that love, which makes for seeing things as they really are, is possible only when one is alone, this sounds very much like what Krishnamurti often said as regards aloneness of the mind. What has to be asked, however, is whether De Mello's aloneness is that which results when self withdraws from others, from things. It seems that it is, and, if it is, this is not Krishnamurti's mind alone. One is not really alone if self is there, that is, there is no state of the truly meditative mind if self is there. If self is there, there cannot be that attention which is love, attention which permits really seeing what something actually is. Such attention and the understanding that comes of it is the only action there need be, and this action is not some mere activity which a partially functioning mind might judge to be the proper response which something deserves or calls for.

If there is merely perception on a dualistic mind level, that is, when perceiving self and that which is perceived are there, there cannot be a total, complete seeing of something. Similarly, if a partially functioning mind is imposing silence, there will not be real silence, and, when there is no real silence, no meditative mind silence, there cannot be love. If one never sees non-dualistically, one cannot love things which relate only to a meditative mind, namely, eternal realities and values, and these will not at all inform a way of living that is truly happy and joyful.

1.26
De Mello: Compassion can be very rude ... jolt you ... roll up its sleeves and operate on you.
Commentator: Real compassion comes gently, like a refreshing breeze. It comes to a quiet, tranquil mind on the ocean of life in the way that a sudden, but gentle gust of wind quietly and seemingly without any effort at all fills the jib of a sailboat and moves the craft over the quiet lake water. It blows into an empty room, one with windows wide open, and fills the room with its freshness. Self, even when it is called compassionate, can be rude, but a truly meditative mind is never so, and this is because such a mind is alone, empty of self and of all that self fashions. Only when there is such a mind is there real love, real compassion.

1.27
De Mello: To acquire happiness you do not have to do anything, because happiness cannot be acquired .... Because we have it already.... You have got to drop illusions. You do not have to add anything in order to be happy.
Commentator: Although unhappiness is caused by a partially functioning mind, real happiness, which is not opposite to unhappiness, is uncaused. It is a by-product of the loving intelligence of a truly meditative mind. There is no real happiness if there is merely a partially functioning mind. Self dropping something, including so-called illusions, is not the acquisition of real happiness. Self dropping is merely the opposite of self acquiring, and so any so-called happiness that self acquires when it drops things is not real happiness. Since real happiness, like real love, has no opposite, it is not there when there is mere movement in the house of opposites. Both self dropping and self acquiring are related to an acquisitive mind seeking to acquire things so-called material or so-called spiritual, and, so long as the mind is acquisitive, it cannot be happy, cannot love.

1.28

De Mello: You have got to wake up and then you will suddenly realize that reality is not problematic; you are the problem.

Commentator: When self is there, there cannot be awakening of intelligence, enlightenment, wisdom. So-called "waking up" is merely self moving from one state of partially functioning mind awareness to another. True awakening is related to the realization that there is no real "you" to be or not to be a problem.

An arrogant, unreal "I" or self opposing itself to what others say and do, to their beliefs and values, never understands reality. It merely reacts against the so-called respectable society or against some part of society which is considered respectable or not.

It is a profound error resulting from profound conditioning of the mind to think that there is a noble "I" different from the so-called lower self or ego. So long as "I" is there in any form – lower or higher – there can be no profound, quiet mind awareness. Self getting something judged worthwhile is merely the opposite of self getting something judged worthless.[64] Identifying the "I" with things called spiritual is merely the opposite of identifying it with things called material. In both cases it is illusion to think there is a real "I" which identifies. True happiness, which is associated with meditative mind awareness, with the awakening of the loving intelligence related to such awareness, can only come when there is no "I" at all.

1.29

De Mello: Wisdom occurs when you drop barriers you have erected through your concepts and conditioning. Wisdom is not something acquired; wisdom is not experience; wisdom is not applying yesterday's illusions to today's problems.

Commentator: Wisdom comes when the mind is receptive to being touched by it, and the mind is truly receptive only when it is quiet, when it choicelessly and observerlessly observes, when it is alert and attentive, not when self or a partially functioning mind makes effort, struggles to peal off its conditioning, to rid itself of concepts. Self is there whether it moves beyond concepts or whether it continues to keep company with them, and, so long as self is there, there can be no profound understanding, no deep wisdom.

A change in self's thinking does not bring wisdom. So-called new thinking is necessarily old thinking, related as much to experience and the past as is any so-called old thinking. Wisdom is not opposite to conceptually built walls of conditioning, is not experience related, but neither is it sensitivity of the self – the unreal "I" or the unreal "you" – to persons, happenings, or surroundings.

The writings of Anthony De Mello clearly show his prizing of right-brain thinking. Unlike *Awareness*, many of the other things he wrote are made up of stories, poetry, and other right-brain creations, and these very often show the influence of Eastern thought and thinking on his own Christian faith, a faith, however, which was not always traditional. Some would even say that often it was not sufficiently orthodox. Always, however, his thinking was partially functioning mind related as much as that of a theist like John Paul II is partially functioning mind related. Thus his answers to our two questions are not only not full and complete but are even illusory for fully adequate answers imply meditative mind awareness, insight, and understanding on a meditative mind level.

De Mello would have said that indeed a partially functioning mind can be a truly religious mind but only if there is movement at some point beyond the concepts and judgments associated with what might be called left-brain partial mind functioning, if there is what he called knowing beyond knowing, that is, so-called awareness and understanding which are other than those which relate to dogmas, doctrines, concepts, and the like, if one heeds advice which suggests that, since "abstraction is not life,"[64] one should turn "to the heart," should come "back home, to"[65] oneself. It is illusion, however, to say that "life is found in experience,"[66] that one can "experience reality .... come to" one's "senses," and that "that will bring" one "to the now."[67] Only a mind which fails to realize that there is dualism of time and the timeless, and that real silence, love, joy, peace, and order relate only to a level of awareness which is not consciousness and to a mind which is truly meditative rather than partially functioning in relation to either the brain's right or left hemisphere would wrongly think and say that there is wisdom in some "fairy tale" dubbed "the story of all of us,"[68] that "a story is the shortest distance between a human being and truth,"[69] and that ours is an "essence" which is "unchangeable"[70] and able to think and feel, indeed to know, reality. Only one whose mind is only and ever partially functioning rather than silently alert would wrongly say that deep understanding comes out of experiences, especially those called mystical, that the so-called "spiritual self"[71] is something real, that "life is based on emotion,"[72] that "emotion is the basic rule of life,"[73] and that real love is about showing "feelings to others."[74]

The truly religious mind, it must be said, is associated neither with the right-brain partially functioning mind of someone like De Mello nor with any theistically functioning mind which relates to a prizing of left rather than right-brain thinking. As regards the second question, De Mello would be not unlike those contemporary thinkers who strive to bring science and mysticism together, a move no better, however, than that made by a partially functioning mind to distinguish the two so as merely to promote the one and to deny that the other has any importance. Science has importance, but of ultimate importance is the truly meditative mind, that mind which realizes that any type of mysticism, Christian or otherwise, and any theism, based as it is upon a prizing of reason or intellect in union with faith, do not make for a truly religious mind. Nor does fideism or any other -ism make for such a mind.

In the next chapter we turn to examples of Eastern religion. There we see what answers it might give to the two questions already posed.

Chapter Two

Eastern Religion, Science, and the Meditative Mind

> It is only when the mind is totally still that there is a possibility of the coming into being of that which is beyond the measure of the mind. But organized religions merely condition the mind to a particular pattern of thought.
>
> To meditate is to purge the mind of its self-centered activity. And if you have come this far in meditation, you will find there is silence, a total emptiness. The mind ... is completely alone, and being alone, untouched, it is innocent. Therefore there is a possibility for that which is timeless, eternal, to come into being.
>
> J. Krishnamurti

The quotations which make up part of this chapter are once again in two sets, the first of which represents what might be said to be a traditional Buddhist or Eastern approach to our two questions and the second a contemporary, right rather than left-brain approach. The cited passages in the first set are from *The Teaching of Buddha*[1] and those in the second set are from Steven Harrison's *Doing Nothing,*[2] the jacket of which makes clear that Eastern thought had influence on Harrison's thinking as expressed in his book.

Some of what Harrison writes seems to be, if not a reflection of things Krishnamurti said, at least very similar to things one reads in Krishnamurti's writings. In actual fact, however, the things said by Harrison in this regard are not at all related to Krishnamurti since what one reads in Harrison, as well as things one reads in *The Teaching of Buddha* are totally related to a partially functioning mind whereas what Krishnamurti says is always out of a truly silent, meditative mind. We turn initially to statements taken from *The Teaching of Buddha*.

2.0
Buddhist Promoting
Foundation: Both flesh and spirit are related to conditions and are changed as conditions change.

Commentator: Flesh and spirit constitute a false dualism. Changing conditions relate to the merely temporal and relative. A truly meditative mind communes with the eternal and absolute even when it holistically observes and fully understands what is temporal and relative. Its communion with the eternal and eternal values – love, freedom, peace, joy, beauty – is steadfast even as there is choiceless and observerless observation of time-related changing conditions and the total clarity of perception without the perceiver of what is true and what is false as regards the temporal and the relative.

Another false dualism is that of this world and the so-called other world.[3] There is only one world, and mind in relationship to this world can function partially or function totally, can be quiet, silent, meditative. It is exclusive partial, inappropriate functioning of the mind which causes misfortunes and sufferings in the way lives are lived. Only when there is the total attention of the truly meditative mind can there be observerless observation of such misfortunes and sufferings and the understanding of them which is the only real, actual transcendence of or going beyond them. Only then can there be the coming of things not opposite to them which make for fullness and profundity of living, that is, timeless living, living in the eternal now, in communion with the real, actual love, joy, freedom, beauty, peace, and order which constitute life eternal.

There is only one world and not two, but in this one world there is the true which is the living, the partially true which is intellectual or abstract in kind, and the false created by irrational mind functioning. With living truth the silent mind communes, just as it is the same silent mind which sees, choicelessly and observerlessly, the false in the world. In the very understanding of this false, this silent mind is beyond it. It is the truly meditative mind which fully attends to what *is*, and this very attention is love. When there is real love, love without self in any form, then there can be real compassion for all.

2.1
Buddhist Promoting
Foundation: There is no failure in Enlightenment; the failure lies in those people who, for a long time, have sought Enlightenment in their discriminating minds, not realizing that theirs are not true minds but are imaginary minds that have been caused by the accumulation of greed and illusions overlaying and hiding their true minds.

Commentator: There is only one real mind, and so rather than talking about imaginary and true minds, one can talk about partially functioning and truly meditative minds. Such talk rightly allows for proper partial mind functioning but also indicates realization that the deep or profound level of the mind is meditative. Enlightenment relates to the totally functioning mind and not to the merely partially functioning one, not to a so-called true mind opposite to a so-called imaginary one. Greed and ambition, moreover, must cease and not merely be struggled against by a partially functioning mind if there is to be the state of the truly meditative, totally functioning, truly enlightened mind.

Narrowness of mind relates, not to a truly meditative mind, but rather to one which merely believes, is opinionated, and is constantly engaged in never-ending partial functioning. A wide-ranging, expansive mind entertains no belief at all of any kind as regards matters of living or relating. Only in the state of the truly meditative mind can there be great wisdom and compassion.

The outer is as the inner. What the mind of man is the world is. When there are activities of the mind resulting from speculation, imagination, and desires related to greed, ambition, anger, and so on, indeed when there are any activities at all, whether ignoble ones or so-called noble ones supposedly leading to enlightenment, the mind is engaged in futile, useless endeavours. Mind engaged in any partial functioning cannot give one a truly quiet mind. By means of time and the known one cannot go to the eternal, the unknown; thus, so long as the mind is engaged in exclusive partial functioning, there cannot be real enlightenment of the mind but only that so-called enlightenment which is illusion projected out of a partially functioning mind. Time, all knowing, and all so-called knowing by a partially functioning mind must cease before there can be awakening, enlightenment, a truly meditative state of mind. The minds of most people are always partially functioning and never silent, and so their minds always and only move up and down the aisle of opposites – good versus evil, like versus dislike, and so on. When this exclusive partial mind functioning is inappropriate or irrational, then people suffer. Clarity of a totally functioning mind, of a truly meditative mind, however, never comes as a result of some partially functioning mind activity, is never, that is, something effected or produced by self at any level, for example, a partially functioning mind or self giving up attachments, striving for clarity of mind, and so on.

Talk of "false" discrimination by the mind[4] implies that there is true discrimination, but all discrimination, one must realize, is of a partially functioning mind. Discrimination does not at all relate to choiceless, observerless observation, the only kind of observation which allows for total truth, deepest understanding, and complete freedom of the mind. Rather than talking in terms of a false dualism of the so-called real or true mind as opposed to a so-called imaginary or "temporary" mind,[5] one can talk about the one mind which can function and which can be silent.

So-called good and evil, so-called love and its opposite, hate, are time-related, but nevertheless real, that is, temporarily but not absolutely real. It is a false dualism of appearances and reality which relates to talk about the essence of mind behind worldly desires, wishes, and wants.[6] Such a dualism is not there when there is talk out of a truly meditative mind about the one mind which can function partially or fragmentarily and the same mind which can be completely silent, which can function totally.

A partially functioning mind is not a mere "appearance" of the mind of man.[7] It is, rather, the one mind of man which is partially functioning, not in a state of total silence. Corruption of the mind is there only when it functions inappropriately, when it functions without due respect for the proper limits of partial mind functioning.

When there is talk about the true or higher mind exercising mastery over the lower mind, this suggests mind control which is said to be necessary for tranquillity of the mind. Whenever control is there, however, the mind is functioning partially, in relationship to a mere fragment or part of the mind. The very presence of the fragment which is the self and its activity or partial functioning preclude the state of the truly meditative mind.

Any idea and all ideas are related to a partially functioning mind. Thus to talk about ideas of a so-called true, pure mind which are not, like some other ideas, "perverted"[8] is wrongly to confuse a partially functioning with a totally functioning mind, and the temporal and relative with the eternal and absolute. To talk about a discriminating mind as different from a "pure mind"[9] is to subscribe to the false dualism of appearances and reality. Though there is the one real dualism of the temporal and the eternal, there is only one mind which can discriminate or engage itself in some other similar activity, or which can be totally silent. The mind that discriminates is the true mind for there is only one mind, but, when it discriminates, it is not meditative. Sometimes, when it discriminates, or attempts to discriminate, the mind functions inappropriately. This it does when it sets up merely imagined, mind-projected differences which are not real differences. By its improper and inappropriate functioning the mind can produce suffering. When there is the total understanding of the truly meditative mind, however, there is mind functioning partially in relation to time and the measurable only when there is necessity for the mind so to function, only when it is appropriate that it so function. Such proper partial mind functioning is never something opposed to the meditative mind's loving communion with the absolute, the eternal.

2.2
Buddhist Promoting
Foundation: We cannot say that an inn disappears just because the guest is out of sight; neither can we say that the true self has disappeared when the defiled mind which has been aroused by the changing circumstances of life has disappeared. That which changes with changing conditions is not the true nature of the human mind.

Commentator: There is no real, true self, but rather only that which is projected out of a partially functioning mind and called real or true. To talk about a so-called true or real self as different from ego or so-called false self is to talk the language of opposites. Opposites are never real unless they are physical like day and night, man and woman. One speaks with full, total awareness which is not consciousness if one talks about the one self, whatever its form, which is there whenever there is partial mind functioning but is not there when there is the state of the meditative mind. Rather than talking about two minds, one which discriminates and is corrupted and another which is pure and clean, one talks rightly when one speaks about the one mind of man which can function rationally or irrationally, but which can also be silent.

Regarding partial mind functioning, it must be said that sometimes partial functioning of the mind, hence presence of the self, even if it is unreal, is necessary, as when the matter is something scientific, technological, or practical. Often, however, such functioning or such a presence is not only unnecessary but destructive, for example when the matter is that which is living, reality, when the matter is communion of the meditative mind with the eternal, the ineffable, the immeasurable, the absolute, when the matter is the love, joy, peace, and order associated with such communion.

2.3
Buddhist Promoting
Foundation: Buddha ... holds the lens of Wisdom before all human minds and thus their faith may be enkindled.
Commentator: Faith is partial mind projection of the unreal. When faith is there, there is not deep mind wisdom but merely that which a partially functioning mind projects and calls wisdom.

Although through faith one can gain entry into teachings,[10] both faith and teachings relate to a partially functioning mind, and, when there is this partially functioning mind engaged in activity, there cannot be that wholeness of wisdom related to an understanding of the whole, the living. To enter into teachings of the Buddha,[11] or of someone else, and to put them into practice is to employ a partially functioning mind, the superficial mind, and not to understand with the living intelligence and insight of the truly meditative mind. Such a mind does not struggle, exerts no effort to free itself from the delusions and miseries of living caused by improper partial mind functioning, but in its truly meditative state it is beyond all such delusions and miseries. It has observed these, choicelessly and observerlessly, with a selfless awareness, and from that pure observation comes the understanding which is the negation of them. Then there has been an emptying of consciousness so that there is only awareness without consciousness rather than a so-called supreme consciousness with self there being more deeply conscious.

It is a partially functioning mind which cultivates habits.[12] It is likewise a partially functioning mind which engages itself in activities of mind, like ideation, reflection, deliberation, and so on, activities which relate to the dead and gone, to the already known and recognized and not to the new and the unknown. Faith, since it is partially functioning mind created, is not at all related to anything real, the living, the eternal, like love, compassion, joy, peace, order.

Faith is merely related to partial functioning of the mind, and, when there is a so-called "depreciated" self[13] rather than an ennobled or enhanced self, self is still there. The presence of self in any form means there is no silence of the mind, and without silence of the mind there is no depth of insight and understanding but merely life lived very superficially, in accord with merely sensate values, and not at all in relationship to eternal ones.

Emancipation of the mind, that is, total freedom of the mind relates to what is beyond all fragments which are parts of the partially functioning mind – reason, will, faith, imagination, reflection – and beyond all fragments created by such a functioning mind, including anything faith-related.[14] Anything which is partial or fragmentary, that is, of a partially functioning mind is superficial. Only a truly meditative mind is a deep mind, mind totally alone, completely uninfluenced, absolutely unconditioned, and so it is not a mind led by anyone. Such a mind is beyond all power, all authority.

Faith, hope, courage, and charity are merely virtues of a partially functioning mind, not at all related to that real virtue or goodness which is not of the level of thought but rather of the level of the living, that which touches only a truly meditative mind. That which one seeks to attain – be it hope, peace, love, and so on – is always projection by self, by a partially functioning mind, hence not truly living, not real.

A so-called Buddha nature, that is, a so-called higher or deeper self, and faith in it are nothing real. Rather they are mere fashionings or creations of a partially functioning mind. Faith, just like every form of the self and every form which the self creates, is of a partially functioning mind, and how can that which is of the partial or the fragmentary lead to that which is total or whole? A partially functioning mind can never lead to a quiet mind. Only when a partially functioning mind ceases all of its movement or all of its activities can there be the state of the truly meditative mind. To think that a partially functioning mind can lead one to a quiet mind is wrongly to think that there is a path from the known to the unknown, from the relative to the absolute, from the measurable to the immeasurable.

2.4
Buddhist Promoting
Foundation: Let us follow the Buddha's teaching and cross over to the other shore of Enlightenment.
Commentator: There is no real other shore but only one imagined, fashioned by, projected out of a partially functioning mind. To say that there is is to talk in terms of a false dualism. Then, to follow a path, any path said to be noble or not, is to engage the mind in futile activity.

Enlightenment can only be on this shore, and it comes about when the mind stops all of its partial functioning, when it is truly silent and thus able to have real contact with the living, the real, able to commune with values which are real – love, freedom, living truth, goodness, peace, order.

There can be no leading of another or "others to Enlightenment,"[15] to an understanding of living truth, just as there can be no way to it or method of attaining it. There is no path at all to living truth, neither that of thought in some form, be it theological, philosophical, or some other, nor of imagination. Also, compassion is never something which can be practiced by a partially functioning mind but rather is there only when there is a truly meditative mind, that state of mind which is requisite for an understanding of living truth, love without self, total freedom, eternal peace, and absolute order.[16]

This world, perhaps called "the world of delusion,"[17] and the implied other world where there is enlightenment constitute a false dualism. Enlightenment is on this shore, the only shore there is, and it is there when there is total attention to what *is* in the state of the meditative mind, and when there is the insight and understanding which come when there is this attention. There is no way to and no method for enlightenment. All struggle and effort of a partially functioning mind must be ended or must cease if there is to be enlightenment.

2.5
Buddhist Promoting
Foundation: The radiance of ... lotus blossoms brightens the path of Wisdom, and those who listen to the music of the holy teaching are led into perfect peace.
Commentator: There is no path to profound wisdom, wisdom which relates to a truly meditative mind rather than to a partially functioning one. That perfect peace which is peace on the level of a truly meditative mind is not something acquired, arrived at or attained by some process of becoming effected by a partially functioning mind. It is given or comes only to mind which has ceased all of its activities and is totally silent even as it is highly alert in its deep passivity.

"Duality" cannot meld into or evolve to "oneness."[18] All duality, except that of time and the timeless, must cease for oneness to be. Thus there cannot be any way or path to living truth, the total insight and understanding which make for true enlightenment of the mind, the profound wisdom of the meditative mind. All supposedly living related concepts and other concepts opposed to these, whether "prejudiced" or not,[19] relate to a partially functioning mind, and, so long as such concepts are there, so long as mind is not emptied of them by total attention to or observerless observation of what *is*, there cannot be true enlightenment of the mind.

The saint[20] is one whose mind is ever partially functioning in relation to some so-called religious tradition, in relation to one of the organized religions, and thus her or his understanding of things like so-called "noble truths"[21] is not that of mind in a truly meditative state but rather is merely an intellectual or supposedly intellectual understanding on the level of a partially functioning mind. Any so-called path to enlightenment related to such so-called noble truths has nothing to do with the understanding of living truth and a life lived out of that understanding but rather has to do with a partially functioning mind which struggles or exerts effort to follow the path which it itself projects as a way or a road to enlightenment.

That which is projected by a partially functioning, imaginative, wishing, and wanting mind is not anything real but rather that which is false, illusory. When there is so-called truth, and a way and method to it which is taught, the teaching[22] involves a teacher, that is, a supposedly knowing self, one with a partially functioning mind attempting to pass on to the taught partially functioning mind fashioned content, partially functioning mind accumulations, and such accumulations are all dead things, related as they are to measurement, past experiences, memories, time-related images and concepts. This teaching, therefore, can say nothing at all related to the new, the unknown, the absolute, the immeasurable, those things of which there must be awareness without consciousness if there is to be true awakening or enlightenment of the mind.

If one says that there is a path to enlightenment, even if it be the one related to so-called understanding of so-called noble truths, if one says that one can use a raft or vehicle, big or small, to go from this shore to a mind-projected other shore of supposed enlightenment, one is in effect wrongly saying that one can go from the known to the unknown, from the measurable to the immeasurable, that one can use a partially functioning mind to acquire or get a truly meditative one.

When there is the flight of the alone to the alone, hence a flight without any guide to offer aid or help,[23] when, this is to say, there is a meditative mind in communion with living truth, with living reality, this has not come about as a result of a process involving struggle and effort on the part of a partially functioning mind or self to transform itself into a so-called truly meditative mind or self. A really meditative mind is there only when a partially functioning mind ceases all of its functioning, all of its many activities to become, grow, fulfill, enhance itself, when it becomes completely silent, when it is highly alert, totally passive, but not passive as opposite to active on the level of a partially functioning mind. The state of the truly meditative mind – a state that is there only when there is no self, hence no desiring and no consequent suffering – is not something to be achieved or arrived at by struggle or effort. There is no path to it, no leader or master, nor any so-called authority who is said to know but really does not know, that can lead one to it, but rather it is there only when there is an emptying of consciousness by choiceless and observerless observation, by attention without an attend-er, and when there is the understanding which comes with such observation, such attention. Those who seek enlightenment[24] will never find it or will find only what is imagined or projected to be enlightenment but which is not at all true enlightenment. Real enlightenment relates to a truly silent, alertly passive mind, mind utterly, totally unconditioned by anyone or anything, not even conditioned by so-called personal or so-called "lived" experiences.

2.6
Buddhist Promoting
Foundation: Quiet meditation distinguishes the seeker for Enlightenment.
Commentator: When there is meditation with a seeker, this is meditation with a meditator or self present in the activity which is called meditation. Such meditation is not of a truly silent mind, the only state of mind which is truly enlightened, fully awakened. Likewise, when self transcends, when it either detaches or attaches itself,[25] or engages itself in any other similar activity, its state of mind is not truly meditative. Self's so-called transcendence is not real transcendence for it is related to effort of self, of a partially functioning mind to seek out what it itself projects, namely, something unreal, abstract, something merely imagined and desired. Attachment or detachment is also of a partially functioning mind, and to be detached from so-called worldly things and attached to so-called spiritual things is still attachment on the part of self, of a partially functioning mind.

"Delusion and suffering"[26] do not derive from the dynamism and changeability of that which is living, but from a partially functioning mind which, rather than directly facing the living in all its dynamism and richness so as fully to understand it and to live in communion with it, escapes from it, creates the illusory, the merely imagined, the unreal world of ought and should, a world merely wished for, longed for. Futile is all effort by a partially functioning mind or self to attain this unreal world – a world for which there is ceaseless struggle so that there might eventually be a so-called fully satisfied self, a so-called fully developed and fulfilled person. All teachings and thinking[27] related to partially functioning mind activity and all so-called help of a supposed master or authority can never make for a truly enlightened mind, that level of mind beyond the level of partial functioning. There is no deep thinking. That which is deep is the loving intelligence of a truly meditative mind.

To throw away some things and to seek for something else[28] is to engage a partially functioning mind. This engagement is in a futile attempt to employ what is partial and fragmentary, that is a superficially functioning mind, to get, to acquire, to attain to a holistically functioning, meditative state of awareness.

Happiness[29] comes not on the level of conscious awareness, that is, it is not something which can be directly achieved or acquired by a partially functioning mind or self engaging itself in effort and struggle, striving to become something merely mind-projected, hence of the unreal rather than the real, that is, something wished for, longed for, imagined, speculated upon, and so on. Happiness comes to the mind when it is in a truly meditative state, in a state of awareness which is not consciousness, a state wherein there is no self there being conscious, no self there which is the knower knowing the known. Happiness is a by-product of choiceless and observerless observation on the part of a meditative mind, mind which understands profoundly and totally. It relates to the profundity, richness, and significance of a way of living actually or really related to the eternal and its values – love without self, total freedom of the mind, and absolute tranquillity and order.

2.7
Buddhist Promoting
Foundation: [In his love, Buddha] conceives the wisest methods to lead, teach, and enrich [people] with the treasure of Enlightenment.
Commentator: Methodologies of every kind and conception relate to a partially functioning mind, to that level of mind on which there cannot be enlightenment, that is, fullness of awareness, insight, and understanding. All activity of a partially functioning mind must cease if there is to be enlightenment. Also, there can be no leader, no teacher to assist another to achieve enlightenment. Enlightenment comes only to mind in a truly meditative state, that state of mind which transcends all

effort and struggle, all opposites, all activity of self, for example of the teacher separate from those taught, of the leader separate from the led, of the observer separate from the observed, of the knower separate from the known. Utterly futile is effort of the self when it seeks to know the unknown, to go from the temporal and the relative to the eternal, the absolute.

No one can give another self-knowledge, that which alone brings freedom beyond psychological fears and sufferings. Enlightenment of the mind is never something which can be attained by a partially functioning mind but rather always something which comes only to a truly meditative mind. Determination to strive for enlightenment and engagement in activities intended to lead to it[30] are activities of a partially functioning mind, and all activity by such a mind can never give one a truly silent, totally understanding mind. It is the same kind of activity which is related to the projection of something after physical death, something unreal, a supposed heaven or hereafter,[31] a reincarnation or a resurrection of some kind.

There is no method, no way, and no system which can lead to enlightenment. To follow some method, some way, or some system which supposedly will lead to enlightenment is to stay ever in conflict and confusion, is to engage in futile activity, is to be in illusion and forever cut off from an understanding of living truth and from life lived in accord with such understanding.

2.8
Buddhist Promoting
Foundation: Buddha-nature ... becomes covered over by the dirt and dust of other interests and people think that they have lost it, but a good teacher recovers it again for them.

Commentator: This statement suggests that a master, an authority, one who supposedly knows, can give or lead another to a truly meditative, enlightened mind. One who says he or she knows the living does not really know it for there can be knowledge in the ordinary sense only in relation to the dead, that is, to the abstract, the inert. Self-knowing alone relates to awareness on the level of the meditative mind, but this knowing is not knowing in the usual or ordinary sense, that is, in relation to images and concepts, accumulations on the level of a partially functioning mind. Self-knowing is knowing the self moment to moment, is knowing now of the now which is the living, the dynamic and ever-changing.

2.9
Buddhist Promoting
Foundation: A wise man seeks to control his mind so that it will function smoothly and truly.... To conquer oneself is a greater victory than to conquer thousands in a battle.
Commentator: True wisdom is of a truly silent mind and not of a mind which is controlled. One who seeks to control his or her mind, to conquer what he or she calls himself or herself engages in partially functioning mind activity. Such activity relates to the superficial and not to the profound, to the measurable and not to the immeasurable, to the known and not to the unknown, to the temporal and not to the eternal.

Effort[32] always relates to a partially functioning mind and never to a truly silent, meditative one. Effort of any kind, even if constant and unflagging, can never lead to true enlightenment, awakening of the profound level of the mind.

A controlled mind engaging itself in so-called virtuous activity[33] is a mind merely called virtuous. Acquired and cultivated virtue,[34] one example of which is that which is opposite to greed, is not real virtue, not at all related to that real goodness with which a truly meditative mind communes. Total, unlimited freedom is only of a mind totally uninfluenced, completely unconditioned, and thus it can never be there when there is a partially functioning rather than a totally functioning, meditative mind.

There is no real self at all, to be mastered or to be controlled but only that self which is a mere creation of thought. Only when there is no self and nothing which the self produces is there total freedom of mind alone, mind truly meditative, totally unconditioned, uninfluenced, in communion with the real, the eternal, the living, the immeasurable. The state of the really meditative mind is not there when there is self which constantly chatters,[35] which ceaselessly busies itself in a few or in many different activities. Only when all talk and all effort by the self ceases is there the truly meditative mind.

If the problem[36] is a living one, having to do with living, with relating, that is, and not a theoretical, abstract one, what is appropriate is full attention to the problem, attention related to observation out of a non-dualistic state of mind and not to a dualistic state of mind with its observer facing the observed. Moral training and controlling the mind[37] can never put one in direct contact with such a problem. Only a silent mind can be intimate with such a problem, and therefore only such a mind can fully understand the problem. This understanding is the answer to the problem, is the answer which is in the problem, and it alone is the real, definitive solution to the problem. Specification of terms which are said to be necessary for the acquisition of enlightenment and any movement of mind resolved to reach enlightenment indicate engagement in effort to control the mind.[38] To engage in a process of thinking which leads one to conclude that silence of the mind is of value[39] indicates activity of the mind, and, where there is any activity of mind, there can be no real silence of the mind. To determine that one must acquire and abide in silence of the mind is to do that which prevents the mind from really being silent. To have a mind filled with teaching of any kind[40] is to have a mind filled with dead things, however much effort might be made to suggest that these dead things are properly called noble, sacred, profound, or wise, and such a filled mind is not open to the new, the unknown, the fullness of what is living. Only mind in a meditative state is fully alert, truly wise, profoundly silent, and completely understanding.

It is self which can be or become humble or vain,[41] but it is also self which futilely seeks for what is unreal, what is merely projection out of itself. It is likewise the self which, in its confusion, exerts effort, judged by itself to be sufficient or adequate. A busy mind, as much as an idle mind, engages itself in futile activity when it strives for the unreal, the merely desired, the merely longed for.

Self is there when there is effort to control the mind, when the mind is judging or evaluating, when it recalls that which is dead, namely, teachings which are mere accumulations fashioned by a partially functioning mind, whether these be those of the Buddha and related to "goodwill" and so-called "kindness"[42] on the part of some self, some partially functioning mind, or some other teachings, perhaps those which form the basis for some organized religion other than Buddhism but which, just like the Buddhist ones, are not at all related to a truly religious mind, that is, a really meditative mind. Any thoughts which might be called sapiential and selfless are merely the opposite of thoughts called foolish and selfish, but all thoughts relate to a partially functioning mind and not at all to a truly silent one. Only a really silent mind is related to what is deep and profound, the living, rather than to what is dead. Only it is related to what is real, absolute, and immeasurable rather than to what is ultimately unreal or merely temporarily real, relative, and measurable, including fashionings of a partially functioning mind which have to do with envy, ambition, greed, and so on.

It is a partially functioning mind which would suggest that one should cultivate good habits to counter bad habits.[43] Such cultivation and opposition, as well as believing of any kind, both "believing in" and "believing that," and the level of understanding to which belief and believing relate are all partially functioning mind related, and in the state of a partially functioning mind there is not, and cannot be, direct contact with what is really good, what is living, what is of the active present, of the eternally now moment – the eternal and all of its values, namely, love, joy, peace, order.

Real virtue or goodness related to the real, the living, is never what can be accumulated by a partially functioning mind for it is of the living and not the abstract, of that which is dynamic and in constant change and not static. Any projection by a partially functioning mind, like a so-called compassionate or eternal Buddha,[44] is a creation of a partially functioning mind, hence related to the unreal, the imagined, the merely conceptualized, the merely ideated, and not to what *is*, that with which the truly meditative mind communes.

2.10
Buddhist Promoting
Foundation: Men do not realize that by the practice of spiritual concentration of mind, they can subdue all worldly desires ... the pure disciplines of thought and conduct ... will carry them beyond all desires and arguments.
Commentator: Any concentration of mind, even if related to so-called pure disciplines of thought and conduct, is a limited and limiting activity of the mind, and practices of any kind can only be related to such activity of the mind. To subdue or sublimate desires is to employ a partially functioning mind as a way to getting a quiet mind, is to seek to go from the known to the unknown, from the measurable to the immeasurable. Any transcendence related to such activity of a partially functioning mind is not movement to the real but rather to the unreal, the merely imagined, what is said to be the ought or the should, the merely wished for, perhaps the longed for.

Self is there when there are feelings of guilt, upset, and so on. It is self also that would determine not to feel guilt, upset, and so on. When self is there, there cannot be actually wise living in the eternal moment, in the active present.[45] When self is there, there is only dwelling in the past and/or dwelling in the future. Only the truly meditative mind lives in the eternal now with mind functioning partially in relation to the past and the future only when it is necessary or appropriate for it to do so. Concentrating "the mind on the present moment"[46] is mere narrowing down activity on the part of a partially functioning mind, and, when there is only a concentrating mind, there is not mind aware of the eternally living, not mind expansively aware of the living in all its dynamism and richness.

To concentrate the mind is to narrow down the mind, to conflate it. To nurture faith, to cultivate it, or to seek to perfect it is likewise to narrow down the mind to a mere fragment or part, a fragment or part which by its very nature is limited and cannot observe what is living, except fragmentarily, partially, and often illusionally in terms of bias, prejudice, distortion. Living truth, truth as related to what is living, of the eternally now, cannot at all be grasped or captured by a fragmentarily functioning mind. It comes only to a totally quiet, meditative mind. Any so-called awakening of faith[47] is related to a partially functioning mind activity, and all such activity can never lead to enlightenment. It can lead only to an entertaining of some mere projection out of such a functioning mind, of the unreal and not the real, of the dead and not the living.

Following "precepts" and practicing "concentration of mind" can never make the mind profoundly wise.[48] These relate to mere partial functioning of the mind. When there is such following and concentration, self is there following, concentrating. Mind engages in futile activity when it struggles or exerts effort to transform its partial functioning into total functioning, when, that is, it exerts effort to become quiet rather than ceasing all activity and thus actually, really being quiet. Such futile activity is of what can only be properly related to the known, the measurable. Such activity is related to the partial foolishly trying to relate itself to the unknown, the immeasurable, the whole. The activity is of self exerting useless effort to a merely mind-projected end.

2.11
Buddhist Promoting
Foundation: One who is to enjoy the purity of both body and mind walks the path to Buddhahood, breaking the net of selfish, impure thoughts and evil desires. He who is calm in mind acquires peacefulness and thus is able to cultivate his mind day and night with more diligence.

Commentator: Body and mind are a false dualism, related to the separation of the one human being, a psychophysical organism, into partially functioning mind projected parts or fragments. There is no path to enlightenment of the mind, and enlightenment can never come to the mind so long as there is self, perhaps called deeper or higher, striving to quash certain thoughts and desires called selfish, supposedly different from other thoughts and desires which are not selfish, hence of a so-called good rather than a so-called evil kind. All thoughts and desires are of self, of a partially functioning mind, of a conscious mind which entails observer, that is, subject or self separate from the observed, from the object. Partial functioning mind quietening of the mind does not make for mind which is truly quiet, silent. True peacefulness is not something which can be acquired by a partially functioning mind effort to become peaceful. So long as there is concerted cultivation of anything, for example, of virtue or of something else related to partial mind functioning, there cannot be the state of the truly meditative mind. All activity or all movement of a partially functioning mind must cease if there is to be enlightenment. Enlightenment comes to the mind; mind can never go out to seek, to struggle, to achieve, or to get it.

There is no way to enlightenment, and to practice a way which one thinks will lead to enlightenment is to engage oneself in futile activity. To so-called see or to believe in anything supposedly real or living, but which actually is merely a projection out of a partially functioning mind is to see or to believe in the illusory, the false. If one says that the eyes which see before enlightenment are the same as those which see after enlightenment, that the mind which believes before enlightenment is the same as the mind which believes after,[49] this could wrongly be taken to suggest that one can go by way of the known to the unknown, that one can go via the measurable to the immeasurable, when in actuality there can only be enlightenment of the mind when observation with the observer ceases, when there is an ending to all forms of belief, all "believing in" and all "believing that." When there is such cessation, such ending, there is then only observerless observation and mind without any belief at all, both of these being prerequisites for mind in a truly silent state. Though there is only one mind, there are different states of the mind, some which involve awareness which is consciousness and some awareness which is not consciousness. The one mind of man can engage itself in activities or it can be completely silent. It can function partially or totally, can be in movement or can be totally silent, without any movement at all, or with a unique kind of movement not at all related to that of a partially functioning mind.

What *is* or reality can be known only in the state of the truly silent mind. What the world is can be totally understood, and, when, out of a meditative mind, the false in the world is observed for what it is, it can be gone beyond. Such a going beyond, such transcendence, however, is possible only when there is a really meditative mind. When there is this state of mind, there can be choiceless, observerless seeing of the true as the true, the false as the false, and the true in the false. Such seeing is with no see-er seeing, no observer observing, but rather there is just seeing, just observation.

When there is seeing from out of a truly quiet mind, then there is understanding that the falseness in the world and the good projected out of a partially functioning mind as something opposed to this falseness are not ultimates. Ultimate good cannot come out of evil. Evil must cease or be ended for ultimate good to be there, just as falseness must cease or be ended for living truth to be there. Ultimate good, the whole, is the eternally living, not something which can be divided up as a so-called good versus an evil on the level of a partially functioning mind. The illusions of the world which make for falseness in the world are temporal, relative realities, not at all related to the eternal.

Not only can "ignorant people ... not know the truth concerning the world,"[50] but neither can any so-called saints or intelligent people, people whose minds are never totally silent, never truly meditative. Even minds which are truly silent do not know but rather are aware with awareness which is not consciousness of what living truth is, can be aware of, can know moment to moment what is true and what is false in the world.

It is a partially functioning mind which cultivates virtue.[51] True virtue is not anything to be cultivated but is there when there is the love and the goodness with which only a truly meditative mind can commune. Beliefs also relate to a partially functioning mind whereas mind in a meditative state is beyond all beliefs related to matters of living, of relating. All such beliefs are necessarily divisive to a greater or lesser degree. A truly meditative mind does not approve of or affirm the world, nor does it negate it in the ordinary sense of negation, that sense which relates to a partially functioning mind. Rather such a mind looks at the world, choicelessly and observerlessly, and thus it sees and understands it for what it is. It is this very understanding of the world, of its falseness, of its illusions, which is silent mind negation of it.

The commentary on these passages from the Buddhist Promoting Foundation make it very clear that Buddhists believe that it is possible for a partially functioning mind to achieve quietness of the mind, for such a mind to become a truly religious mind. Professor S. Rinpoche indicates that this is so when he writes that "among Buddhist people thought has been accepted as one of the means or methods during preparation"[52] for transformation or enlightenment of the mind. When, however, Rinpoche says simply that Krishnamurti "does not accept or does not talk about preparation,"[53] this does not put the matter rightly. What needs to be said, but which is not said, is that all means and methods to attain to enlightenment are futile or useless. This Krishnamurti saw out of a truly meditative mind, I would suggest, and, when there is seeing on this level of the mind, there is neither acceptance nor rejection of anything. To accept and to reject something is to be engaged in a partially functioning mind activity, and, since, as Rinpoche says, Krishnamurti "never spoke at the relative level" but "always spoke at the level of the absolute,"[54] it is not totally correct to say that he "never accepted" any "graduations and methods."[55] He neither accepted nor rejected them but saw their complete uselessness as regards mind experiencing that fullness of awareness which is possible only when there is true meditation.

When Rinpoche rightly says that "the Buddha adopted" thought processes "as one of his methods to help people take a deeper inquiry,"[56] he does not say as he might that any such process to attain enlightenment relates to futile activity and effort. Because Krishnamurti saw the futility of any such activity and effort, saw the impossibility of any thought, of any partial functioning of the mind to effect enlightenment of the mind, Rinpoche is wrong when he suggests that Krishnamurti would have said that a thought process is simply "a longer journey" to "the absolute truth."[57] What Krishnamurti so often says is that there is no way, no journey, no path to living truth, to enlightenment of the mind. Furthermore, if Krishnamurti was silent or did not speak of those "graduations and methods"[58] which Buddhists and others say can prepare a person for enlightenment of the mind, this was because, it can be suggested, he saw the futility of all such preparation, of any so-called master or authority striving to help another to transform himself, to provide assistance as regards a process believed to lead to enlightenment. Any system, method, way, measure, or guru, Krishnamurti made clear, can only militate against real revolution in the mind, real enlightenment of the mind.

To say that all methods and means, all authority and systems are futile as regards the acquisition of a quiet mind is correctly to answer the first of our two questions. A partially functioning mind, this is to say, can never become a truly meditative, really religious mind. To say that science is limited and necessarily related only to partial functioning of the mind and never to a truly meditative, religious state of mind is correctly to answer the second of our two questions. We turn now to what was earlier called an example of a contemporary, Eastern approach to these two questions, an approach perhaps not incorrectly described as more right-brain than left. In any case, this approach is that taken by Steven Harrison in his book, *Doing Nothing*, mention of which was made earlier.

2.12
Harrison: [Advice is being offered when one entitles a book]: *Doing Nothing*.
Commentator: A careful reading, out of a truly meditative mind, of the book jacket for *Doing Nothing* brings realization of the illusory ideas there expressed. To talk, for example, about a "map to that space ... where divine consciousness is active," to say that this book "offers seekers a way to touch the truths of life," and to refer to Harrison's voice as being "a welcome companion on our journey toward being fully human" and to "Nothing" as the "active place" where "we discover who and what we are"[59] are all examples of such illusion, such falseness.

If there is a map of any kind, and a journey related to this map, then there is a way or a path to, and perhaps a method which can be employed to grasp "truths of life," that is, living truth, reality, to attain fullness of insight into and complete understanding of life. Such, however, relates to impossibility. If a way is followed, then there is self on the way. If self or a partially functioning mind stops the search and does nothing, as Harrison seems to suggest it should do, this is not the same as a meditative mind seeing the futility of search and so being quiet. Is the "doing nothing" which Harrison advises, one might rightly ask, an activity of a so-called meditative, contemplative, reflective Higher Self, of the so-called Atman? It seems that it is, but, if self in any form, higher or lower, is there, there cannot be real silence of the mind. Rather there is implied effort to stop doing one thing and to begin to do another. Effort of any kind to reach a truly meditative mind militates against there being a really meditative mind, the only state of mind wherein there can be a full and complete answer to questions about right living, about right relationship.

The "space" spoken about on the jacket to *Doing Nothing* is merely a projection out of a partially functioning mind, and the divine or the sacred there mentioned is not that which is transcendent to all consciousness and thus is not anything real. It merely has to do with awareness which is consciousness and thus must necessarily be of the relative, the temporal, the known, the partially functioning mind.

Real self-discovery is not a matter of some supposed real self discovering "who and what" it is, but rather relates to the truly quiet mind seeing, observerlessly and choicelessly, what the self is, that it is nothing real, and that what it fashions or makes is likewise nothing real but merely some projection out of itself.

2.13
Harrison: [The subtitle of the book is]: *Coming to the End of the Spiritual Search.*
Commentator: If searching by the self in any form, higher or lower, is ended by the self, this is merely self projecting an ending. Only if the ending is related to choiceless, observerless watching out of a meditative mind is the ending really an ending. After "long" and many thought-related experiences, we are told,[60] Harrison reached the goal of his much protracted spiritual search. It is self which sets a goal, which pursues a goal, and which reaches a goal. Self is still there at the end of a search, perhaps imagining that it is fulfilled and satisfied because the goal has been reached, but, so long as self is there, or continues to be there, there really is not an ending. Only when there is a real stop, a total ending of self, of a partially functioning mind and all its creations, is there the state of the really meditative mind, the really religious mind.

2.14
Harrison: Religion reflects something profoundly important .... It is the exoteric formulation of what is hidden inside us.
Commentator: A reflection of something real is not the real. To be aware of the real as the real, to know the real, only possible moment to moment, there must be an ending or a negation of the reflection. If this is so, and religion relates to what is unreal, namely ideas, thoughts, and rituals, then only the truly religious mind, which is the really meditative mind beyond all partial mind functioning, relates to the real.

Religion as we know it is thought, hence not the real. There is no truly, really religious mind unless there is an ending of religion – its dogmas, its rituals, its so-called mystical aspects. Thus, religion is not the beginning of enlightenment. To say it is such is to suggest the possibility of going from the measurable to the immeasurable, from the known to the unknown. Such a possibility is also suggested when Harrison advises going deeper than the exoteric. Exoteric and esoteric are a false dualism for the inner is as the outer and the outer is as the inner. One must negate both the exoteric and the esoteric aspects of religion, must negate all religion in its ordinary senses if there is to be real understanding of the truly religious mind. Rather than a partially functioning mind going more deeply, there can be realization of the significance of deep awareness which comes from observerless looking at what is false, for example, ordinary, so-called exoteric religion and so-called esoteric forms of religion, awareness which comes from the understanding of what they really are and which makes possible a real going beyond them.

Harrison speaks about great religious leaders as people who opposed the religion of their day.[61] It is a partially functioning mind, like that of these so-called great leaders, which opposes something, just as it is such a mind that strives or struggles to break through things which it decides to oppose. If striving and struggling were there on the part of these so-called great leaders, what was also there was an image or an idea projected as the transcendent ideal which they desired to make incarnate and not what is the truly, actually, really transcendent. Their discovery was partially functioning mind and not quiet mind related. If this were the case with these so-called great figures, then theirs was not immediate and direct communion with what is living, what is reality, but rather with something projected out of their own partially functioning minds.

2.15
Harrison: Without acting from a center, everything we touch, everything we do, is changed ... quiet becomes the milieu in which we function .... Our actual requirements are really just to be still, to love, to relate.
Commentator: This statement seems to suggest that there is a deeper, real self which can be still, love, and relate, and that this deeper self is other than the so-called lower self which is active rather than still, which does not love, does not relate. Deeper or higher, and lower self, however, make up a false dualism. Self is there whenever there is touching and doing, whether before or after so-called change. To talk about "becoming" suggests a process of going from the limited to the unlimited, from the known to the unknown, from the recognized to the unrecognized, from the experienced to the unexperienced, from the relative to the absolute, from the temporal to the eternal. The unlimited, the absolute, the eternal, however, can only be there when the limited, the relative, the temporal ceases, is ended, when there is negation of it by mind in a truly meditative state. Without such cessation, any silence which might be said to be there is merely mind-induced or contrived silence, hence not real silence. It is not the silence, this is to say, with which the truly meditative mind

communes, silence which is of the real, and not of the unreal on the level of a partially functioning mind.

2.16
Harrison: Thought is what arises from ... quiet in the field of consciousness.
Commentator: Quiet in the field of consciousness is merely mind-induced silence. It is not real, absolute quiet. It is relative, man made, partially functioning mind produced quiet and not the absolute quiet with which a truly meditative mind communes. There is, however, awareness beyond consciousness to which real quiet relates. Thought is reality in one sense, that is, it is relatively, temporarily real but not at all related to absolute reality, what is living, what is of the eternal now.

Quiet in the field of consciousness is not real, actual quiet but merely that so-called silence or quietness which is opposite to noise, to chatter. It is awareness without consciousness which is limitless. Consciousness and thought are both related to a partially functioning mind, to measurement, knowledge, experience, time, the inert, and the abstract. Only awareness without consciousness is of a fully functioning mind, a truly silent mind, that mind which alone can commune with the immeasurable, the unknown, the eternal, the living.

True silence is not of consciousness. It is beyond consciousness. Consciousness is not without limits. Thought which comes out of it, or rather the thought which is consciousness, the content of consciousness, is related to the limited – the relative, the temporal, the experienced, the known, the recognized, the superficial.

Harrison speaks of the limitations of psychotherapy. He says that after psychotherapy, self, the center, remains and is not transcended.[62] What must be said, however, that is not said by Harrison, is that there is no deeper self which could affirm the nothingness of the lower self, the ego, and cause it to move to a deeper level of consciousness. Indeed, any level of consciousness, even if so-called deep rather than superficial, means self is still there. Only when there is awareness which is not consciousness – that is, truly meditative mind awareness – is there total understanding. For this one must go beyond experience, beyond the psychological pain which Harrison thinks can be a means to the attainment of deep understanding.[63] Such pain, something which is experienced on the level of the psychological, must cease, be negated if there is to be deep understanding of what is living, of the eternal now. To say, as Harrison does, that one must take emptiness to oneself, to embrace it[64] means self is there to do this. Any such embracing of emptiness, of nothingness, is the embracing of what is mere projection of a partially functioning mind and not that real emptiness or nothingness, that aloneness with which a truly meditative mind communes.

Therapy, as Harrison describes it, could only be modification without fundamental resolution. Fundamental resolution is "enlightenment," the possibility of which is not there in Harrison for he dismisses enlightenment of the mind as something merely mythological.[65] Of course, enlightenment is impossible if there is always and only a partially functioning mind, but it is a real possibility in relation to the truly meditative mind.

2.17
Harrison: If ... "me" is a thought form ... arising and passing away as all thought appears to do, then who are we? Who is the observer of this passing away of the "me"?
Commentator: Indeed, "me" or self is merely a "thought form," hence, nothing real. It is, however, identical to the "we" in the question, "who are we?" Furthermore, "the observer" of the passing away is the "me" which passes away. This is to say that the observer is the observed, that there is no real self which sees, observes, watches, and so on. Only when there is this realization, this understanding, does living truth – relative to what is ultimately unreal or merely temporal and to what is actually real – come to the mind.

The only real answer to these questions which Harrison asks which is not illusory is "nothing," absolutely so, actually so. There is no self which is not an idea, hence there is no real deeper or higher self. Self-understanding, moreover, is possible only when there is no self at all, no supposed self which is not idea related and which is said to be undivided rather than divided. Self reduced to nothing, self without identity, is still self, and the nothingness which experiences itself as nothing is still the self even if called higher or non-centered. Self really dies only when there is understanding which comes from attention out of a truly meditative mind, when there is observerless seeing of the self, what it actually is and does. It does not come from reflection, that in which Harrison invites his readers to engage themselves.

Reflection is not real meditation, not meditation focused on the real but rather on what is unreal – the past, memories, experiences. It is self or a partially functioning mind which reflects. When there is quiet mind meditation, there is pure attention, attention without self present in the act of attending. This attention is observation which is non-dualistic, observation when there is no observer separate from the observed. Only an attentive mind deeply understands, and such understanding is what gives insight and wisdom as regards the living of a life which is full and rich, joyful and peaceful. If self is still there after the so-called passing away of the so-called "me," such that so-called space which is silence, emptiness, or nothingness is then there, this space is merely self-projected, that is, partially functioning mind projected space. Such space is not the real, absolute space which relates to a truly meditative, a totally quiet, still state of mind.

2.18
Harrison: There is nothing obstructing silence, and there is nothing to do to find this silence. It is already present, waiting for us to cease trying.
Commentator: What has to be asked here is, "What is this 'us' that will cease trying?" If one takes it as reference to higher or deeper selves, deeper than the lower selves which seek to find silence, one is in illusion. Ultimately, the higher is as the lower, both unreal, both mere creations out of a partially functioning mind.

There is real silence, and indeed effort and striving by self will not give this silence. Silence, however, will neither be there if there still remains self not striving. Self and all its creations and projections must be negated or ended if real silence is to be there.

A so-called aware observer or watcher who is not striving is still self, only supposedly but not really different from "me," from ego. There is true meditation without techniques only when there is just observation or watching and no self or observer at all.

2.19
Harrison: Perhaps it is time to look again.
Commentator: When there is no separation between the observer and the observed, there can be looking related to holistic seeing. If the observer, the look-er, or the self is there when there is looking again, this looking is not holistic, not at all related to depth of insight and understanding completely, not seeing which is other than time-related seeing from memory, from conditioning. Only when looking is observerless is there looking out of a truly meditative mind, the only state of mind related to loving rather than merely rational intelligence, directly related to living, dynamic reality or what *is*, rather than merely to what is dead, static, or inert, to what is merely image, symbol, or idea.

2.20
Harrison: Faith is what is left when what can be stripped away, is. In that, what we come to is not faith in the divine, rather what we come to is the divine.

Commentator: When there is faith, there is the faithful one, the self which believes in or trusts the object of its faith, of its belief, of its trust. Though this self might strip away aspects of its conditioning, other aspects of conditioning still remain. Only when there is a completely unconditioned, uncontaminated, absolutely pristine mind is it possible for the divine or the sacred to touch the mind, but this divine or the sacred is what is real and not merely some projection out of a partially functioning mind. If there is any "coming to" the divine, self is there, and whatever self arrives at cannot really be the divine but rather only what this self projects out of itself as the divine. Faith, either "belief in" or "belief that," are of a partially functioning mind and thus not absolute. Both must be negated in the state of the meditative mind if the sacred, the ultimate, the unnameable, the immeasurable is to be there.

2.21
Harrison: We truly live in this relative world only when we have an intelligence that transcends division .... The amalgam of subjective and objective worlds includes the whole of life, a life into which we merge.

Commentator: Intelligence which truly transcends division is the loving intelligence of mind in a meditative state. Such intelligence, however, understands that the whole of life is not really there when there is amalgamation of "subjective and objective worlds." Putting two "unreals" together does not make the real, but rather the real is there only when "unreals" are negated by a choicelessly, observerlessly watching mind. Unity is beyond any subjective/objective polarity. Life, the living, is not something into which selves can merge, for, so long as selves are there, the whole of life is not there.

As we live in this so-called relative world, we can live fully or integrally only if we are not of it. Only quiet mind intelligence transcends divisions, and it is not at all based on recognition. Recognition is necessarily related to experience, a partially functioning mind, and any so-called unity such a mind might create. All ideal or image relationships, and all self-and-other relationships must cease if there is to be deepest understanding, fullness of intelligence.

Rather than a partially functioning mind making an amalgam of so-called subjective and objective worlds into which self and selves might supposedly merge, an amalgam which would not at all be something living but merely something projected by a partially functioning mind and called life, there can be the actual, living whole with which only a truly meditative mind can commune.

2.22
Harrison: We create ... symbols of unity .... But, because we are still looking within the realm of language and its inherent me-object structure, we never actually come to wholeness.

Commentator: Real wholeness is not something to which a merely partially functioning mind or self, even if called higher or deeper, can come. It is not related to consciousness at all but only to awareness which is not consciousness. Real wholeness comes to the mind which is in a meditative state when there is real communion of a truly silent mind with the living, reality, what *is*. It is not there when a partially functioning mind is seeking or exerting effort to grasp something projected out of itself which is imagined by that same mind to be wholeness.

So long as self looks, whether "within the realm of language" or elsewhere, there can never be wholeness. The experiencing of wholeness is not possible so long as self is engaged in any process of becoming. Indeed, only when self is not there, actually and absolutely, can there be real wholeness, that is, the wholeness which is living and not something merely abstract, hence not real. Real wholeness is not there when self merely stops creating "symbols of unity" as would be the case when self imposes silence upon itself. Imposed silence is not that real silence which is possible only when there is no self at all, no self which would either create or not create.

2.23
Harrison: Enlightenment is a concept, an idea, a belief .... Enlightenment is a myth because the self is a myth.

Commentator: There is indeed no enlightenment if there is only a partially functioning mind which creates concepts, ideas, or beliefs, but enlightenment is the awakening of the totally functioning, truly meditative, religious mind. Self is indeed a myth, but it is possible for mind to be in a state where there is no self and no myth of any kind, and, when it is in such a state, it can commune with the real, what *is*, the living. Such a state of mind is the state of enlightenment of the mind, something real and not at all mythical.

2.24
Harrison: Confusion is the introduction to true intelligence ... without a center and without the dominance of thought.
Commentator: True intelligence is the loving intelligence of the truly meditative mind. Only this intelligence, and not merely rational intelligence, is without self or the "center," and without "the dominance of thought." Indeed, when there is the state of the truly silent mind, there is not only "no dominance of thought," but there is no thought at all. Such a state of mind is possible, however, only when confusion is negated or ended. One cannot use thought or confused thinking to acquire loving intelligence. All of the known, the measurable, the temporally-related, the named, the relative must cease for the unknown, the immeasurable, the eternal, the nameless, the absolute to be.

2.25
Harrison: Fully understanding ... time and memory ... brings psychological conflict into the accessible and immediate present.... To do anything in relationship to the conflict is to give it substance or energy.... If we do nothing, what occurs? Nothing occurs.... This is the resolution of psychological conflict.
Commentator: Is self there "fully understanding"? This is the question which one can ask when this statement is read. If self is there, there can be no full understanding, understanding which would include awareness that the partially functioning mind is not the truly silent, meditative mind. When self does nothing, self is still there, and self which either does something or does nothing is at the root of any psychological problem. When self is there in silence, thus "doing nothing," the silence is not real silence.

There can be observerless, choiceless, selfless observation of conflict by a totally quiet, meditative mind. Such observation gives fullness of understanding and, so to speak, access to the immeasurable, the eternal. This kind of observation is not what is there when there is merely dualistic seeing of psychological conflict and effort on the part of a partially functioning mind to overcome it. Any thought related to the overcoming of such conflict would be partially functioning mind related, and thus any solution to this conflict could only be partial at best if not illusory. Only when there is really quiet mind understanding is there true enlightenment of the mind. When the mind is really enlightened, it is beyond all psychological confusion and conflict. For such a mind there are no psychological problems.

2.26
Harrison: Can we embrace change and understand that it releases us from the endless repetition of thought and conditioning? Change is the only exit from our selves.
Commentator: "Embracing change" suggests presence of the self, of a partially functioning mind. Rather than self embracing change there can be observerless and choiceless observation of change. Because such observation is without self being present to skew or distort the observation, change can be fully understood. Then there can be change which is not mere modification or reformation, but rather true, profound, or radical revolution, with real openness to the new, the unknown, or the living rather than mere focus on and a mere doing related to what is dead and inert. If self is there effecting or striving to bring about change, there cannot be an exiting from self. The only real exit from self which is possible is there when there is the state of the really silent mind.

2.27
Harrison: The recognition of pain is the moment of freedom... we have never fully felt and embraced our pain. If we embrace it, we embrace ourselves, we embrace nothing.
Commentator: Recognition is always of the known, and true freedom is beyond the known. Mind must be beyond the level whereon there is recognition of anything for there to be full freedom of the mind. Self is there if there is an embracing of pain, of nothing, and that which is embraced is nothing real but merely something projected as nothing by self, by a partially functioning mind. If we embrace ourselves, then we embrace what is merely mind-projected nothingness. Actual, real nothingness cannot be embraced. It is there when all effort to embrace, to escape from it, and so on, ceases, when there is no partial mind functioning at all, when the mind is absolutely quiet.

2.28
Harrison: It is only through the dissolution of ... the idea of "love" and the idea of "me" that real love is found. This is not the "love" described by words, but by silence.
Commentator: Dissolution of ideas, by self, could be said to be mere effort of self to move from left-brain to right-brain functioning, to go from idea to image, from reason to imagination so as to recapture what is said to be mystical insight expressed in religious myths. Real, actual love, however, is there only when all activity related to both left and right-brain

thinking ceases such that the partially functioning mind is completely silent with a silence not brought about by any so-called action or reaction which is effort of self. Such love is there only when there is observation of what *is* without observer or self there at all. The very attention given to what *is* when there is observerless observation is real love.

2.29
Harrison: [Contemporary science's] interconnnected universe transformed by consciousness ... is the realization of mystics, [and] each of us has the capacity to connect myth and science, physics and metaphysics, the material and the mystic.

Commentator: Any so-called "interconnected universe transformed by consciousness" is a creation or a projection of a partially functioning mind. So too is any "realization of the mystics." Such a so-called universe is not the actual whole which is real mystery, life itself in its abundant variety and richness. Any connecting of "myth and science, of physics and metaphysics, and the material and the mystic" is a connecting of things which are not real but rather things fashioned on the level of thought, of a partially functioning mind.

Rather than striving to make any such connection, one can see the dualism of the temporal and the eternal. One can see that, while science is appropriately related to the temporal, the truly meditative mind, which is not at all related to science and to mysticism, is related to the eternal. Then there is no fashioning of mere constructs of the mind which are supposedly, but not really, related to deep insight and understanding of reality, of life.

This statement by Harrison clearly indicates that science and mysticism are both on the level of the partially functioning mind. Both are results of partially functioning mind activity. Both relate to the limits of consciousness. Beyond all consciousness and its limits, however, is awareness which is not consciousness, awareness of the truly meditative, religious mind. Profound change is there only when there is such awareness which is not consciousness.

Even if science had that "unified theory of consciousness"[66] which Harrison thinks it does not have but sorely needs, the fundamental issues of life would still not be adequately addressed for beyond all consciousness, hence beyond science and the mysticism which Harrison tries to bring into positive relationship with science, is awareness which is not, cannot be, consciousness. Any more inclusive picture which science might eventually have would never be the living whole with which the truly meditative mind communes.

Real unity or oneness does not come from consciousness. There is integration only when there is awareness beyond consciousness, hence beyond science and mysticism. As Krishnamurti says: "The way of integration is the process of negative thinking, which is the highest comprehension."[67] Such integration is not a combination of so-called mystical insights and the results of scientific thinking. Any such combination could never be a real unity. It could only be something fashioned by a partially functioning mind.

When consciousness strives to be awareness which is not consciousness, the mind can produce myths and metaphysics. Both, however, relate to the unreal, to memory and imagination, to the past and the future as projection out of the past and not to what is of the eternal now, the real, the living. Consciousness is time. It cannot be profound and does not relate to the profound. Because of the one real dualism of time and eternity, there is the partially functioning mind and the truly quiet mind, there is science and the truly religious mind. What is important is to see the difference between time and the timeless, rather than for one to try to amalgamate science and mysticism. So often Harrison suggests that what we need is thought other than linear thought. What we need, he suggests, is more right-brain thinking to counter undue emphasis on left-brain thinking; hence, we need mysticism and mythology and not just logical, conceptual thinking. What we really need, rather, is meditative mind awareness and not just the conscious awareness which relates to the doing of science. There is no need at all, however, for the mysticism and mythology which, like science, relate to consciousness, but which, unlike science, are examples of inappropriate partial mind functioning since they involve a striving, as is not the case with regard to science, to go from the known to the unknown, to measure the immeasurable, to name the nameless.

With these comments, we finish our look at examples of Eastern religion, or of thinking influenced by Eastern religion. As regards the first of our two questions, both the Buddhist Promoting Foundation and Harrison would say that it is possible for a partially functioning mind to come to an understanding of reality, of the transcendent, of the eternal, hence to become or to be the truly religious mind. The former, for example, would say that mind concentration, a choosing of the good rather than evil, movement beyond a merely discriminating mind, cultivation of faith and good habits, and so on, can lead one to enlightenment of the mind, hence to the really religious state of mind. The commentary above on passages by the Buddhist Promoting Foundation, however, makes it clear that its thinking is always on a partially functioning mind level. On such a level of the mind there cannot be any profound understanding. A partially functioning mind can never go by means of the known to the unknown. Disciplining of the mind, experiences, faith, mind control and concentration, and so on, can never bring about a truly religious state of mind. Little, if any, attempt to address the second of our two questions was made in the book which contains the passages by the Buddhist Promoting Foundation cited above.

Harrison makes moves which are similar to those of De Mello, particularly when he suggests that to understand reality there must be movement beyond left-brain partial mind functioning and things which such functioning produces, like ideas and concepts. More prominent in De Mello, however, is talk of a real "I" which is different from ego, of the so-called deeper self, the soul, and this may well be because of some of De Mello's Christian background, because of the theistic religious tradition with which he was identified and which has always greatly emphasized belief in the soul, man as made in the image and likeness of God, and so on. Harrison, even more than De Mello, stresses an amalgamation of science and mysticism. It has already been said, however, that any such move leads to confusion and illusion. Such a move relates to a failure properly to see the difference between time and eternity, the relative and the absolute, the measurable and the immeasurable, in a word, to realize that there is only one real dualism, namely the temporal and the eternal, and the consequent necessity of making a difference between a partially and a totally functioning mind. Such a move also relates to a further failure to observe the proper limits of the partially functioning mind, that is, of reason, hence of science. Science is not what can give a complete understanding of reality anymore than mysticism can. Thus, an amalgam of two "cannot's" – impossibility – cannot give a "can" – possibility. Science is legitimate and necessary, of value within proper limits, those of reason, of logic, but mysticism with its right-brain constructs is not. It, no more than left-brain constructs, can give the insight, meaning, and understanding which are vouchsafed to a truly meditative mind.

In the next chapter the focus is on psychology. Once again the two questions are asked to see what kind of answers are given in books which at least in part can rightly be characterized as psychological in kind.

Chapter Three

Psychology: Science or Meditation?

> The way of religion is the disentanglement of the mind from the pattern which is put together by the collective, by the past.
>
> Without meditation there is no perfume to life, no beauty, no love.
>
> J. Krishnamurti

Within contemporary psychology there are groups or movements which are engaged in futile attempts to bring science and mysticism together. This futility is seen when, rather than there being what Allan W. Anderson refers to as an absolutizing of consciousness,[1] there is realization that, though there is awareness which is consciousness to which science properly relates, there is also awareness which is not consciousness to which alone a truly meditative mind relates. Such a meditative, truly religious mind, however, is not the mind of one who subscribes to any organized religion, nor is it the mind of one who prizes mysticism as a way to acquire a so-called religious mind.

Passages which follow are taken from books by writers who are part of these attempts to unify science and mysticism. These writers absolutize consciousness, and this prevents them from making a radical, but important, difference between a partially and a totally functioning mind. All of the mind, they think, is what can be called a partially functioning mind, and so, were they to speak about a quiet, silent mind, the silence of the mind about which they would actually be speaking would not be real silence but rather merely a silence opposite to talk or noise on the level of the partially functioning mind. When these writers do not make a difference between a partially and a totally functioning mind, they show that theirs is not a realization of the real dualism of time and the timeless, the temporal and the eternal. What they so often say, therefore, is in relationship, not to what is living truth, not to what *is*, not to ultimate reality but rather to what is unreal, what is ideal, what is merely abstract. Therefore what they think and say would constitute inadequate responses or answers to our two questions. First, we listen in this chapter to things said by Deepak Chopra in *Ageless Body, Timeless Mind*,[2] then to things written by Ken Wilber in *Ken Wilber in Dialogue*,[3] and finally to some of what Wes Nisker says in his book, *Crazy Wisdom*.[4] If characterization as "psychological in kind" does not exactly fit anyone of these books, particularly that of Nisker who refers to himself as a generalist,[5] it nevertheless can serve as one more example of the many books which are based solely on a partially functioning mind and on a failure to make an important difference between a partially and a totally functioning mind.

3.0
Chopra: There is something woefully lacking in any fragmentary approach to life, however intriguing any single fragment happens to be.
Commentator: The crucial question here is what constitutes non-fragmentary living. It is not enough merely to recognize a fragment as a fragment but most important to realize a state of mind which is truly beyond each and every fragment. When self, even if called higher rather than lower, or soul, or Atman is there, living is fragmentary. When, as regards living, relating, any level of consciousness – lower, higher, or so-called supreme – is there, there cannot really be non-fragmentary living.

3.1
Chopra: [There are exercises to attain] a level of transcendental experience, which is in fact *more real* than the world of the senses.
Commentator: Sensory experience and transcendental experience relate to the false dualism of this world and another world. There is not, and cannot be any such thing as transcendental experience, but rather only experiencing of the transcendent when mind is in a truly meditative state. The unknown, mystery or living reality, cannot be experienced, and so there is futile activity when a partially functioning mind exercises and

disciplines itself in effort to experience it. Meditation with meditator, with an experiencer there, is not real meditation. There is time to which experience relates, and there is the timeless which, because it is living or the unknown, touches only a really meditative mind, and never a partially functioning one, the one which properly has to do with time, experience, the known, and the recognized. Furthermore, though exercises and discipline might lead to better partial mind functioning in relationship to time – what is wrongly referred to here as some separate "world of the senses" – and to things which are temporally related, all such activity and effort in relation to the real, to living truth, and to a way of living in communion with such truth are futile and illusory.

3.2
Chopra: The modern sage J. Krishnamurti lived into his nineties with wonderful alertness, wisdom, and undiminished vitality.
Commentator: What can be realized and said is that alertness in the case of Krishnamurti, indeed wonderful as it was, was of the level of a truly meditative, really silent mind, and so it was not at all related to the limited, conscious awareness which underpins all that Chopra says and writes. Also, wisdom in the case of Krishnamurti was not so-called wisdom related to consciousness, experience, time, the known, and so on, but rather was the profound wisdom of the truly meditative, completely silent mind. Alertness and wisdom, in the case of Krishnamurti, were not at all related to self-awareness, not to one who might be said to be a "wise person," nor to any so-called discovery of "who one really is," that is, one's so-called deep, inner, but unreal self. They were related rather to the completely attentive mind, which, because of its observerless, choiceless awareness, alone can be profoundly wise and truly loving.

3.3
Chopra: The highest state of consciousness available to us is unity, which erases the distinction between observer and observed.
Commentator: True unity, integration, and wholeness of the truly meditative mind are not at all within consciousness, not at all related to any form, level, or stage of consciousness talked about by a partially functioning mind. Though there can be awareness of mind transcendent to or beyond "the distinction between observer and the observed," this is not effected by self erasing such a distinction. To think that self or a partially functioning mind can do this is wrongly to think that there can be

movement from the known to the unknown, from the relative to the absolute, from the limited to the unlimited, from the measurable to the immeasurable. Such transcendence to true integration or to real unity is only possible when there is understanding by mind choicelessly and observerlessly observing what the self really is, when there is understanding that the self which is something merely imagined or believed to be real and capable of understanding the nature of its own presumed reality is actually not at all anything real.

3.4
Chopra: A definite, conscious process ... extends beyond the individual to a larger reality – for some this is God ... for others it is the Self or the Absolute.
Commentator: Such a statement indicates a failure to realize that there is real dualism of time and the timeless, and that there is the consequent necessity of understanding the difference between a partially and a totally functioning mind. Both the individual and the collective which stands opposite to the individual must be negated or ended if there is to be any so-called larger reality. The larger reality spoken of here, even if called God, Providence, Self, or the Absolute, is not anything real but rather mere projection from out of awareness which is merely consciousness. This unreal "reality" comes out of a mind which is merely functioning partially and is not meditative, which is engaged in mere activity and so is not really in action, which is to say, in love, in communion with the living and not merely abstract whole, with the ultimate, the absolute which is real and not a mere fashioning by a partially functioning mind.

3.5
Chopra: Deep inside us ... is an innermost core of being, a field of non-change that creates personality, ego, and body.
Commentator: Any such "core of being" is a mere fashioning of a partially functioning mind, of a mind which is merely speculating or imagining, hence not at all really related to what *is*, the real, the living. Inner and outer constitute a false dualism, as do the material and the spiritual. Though the body and the mind are physical, hence real, personality and ego are not since they are mere fashionings of a partially functioning mind, of self, each one a creation of thought as much as is the self which fashions them.

3.6
Chopra: What the world's religious traditions call Spirit is wholeness, the continuity of awareness that oversees all the bits and pieces of awareness.

Commentator: The wholeness spoken of here, as much as the so-called Spirit which is said to equate to it, is merely a creation by a partially functioning mind. Awareness which can really "oversee" the many "bits and pieces of awareness" which is consciousness is awareness on the level of the truly meditative mind, is awareness, therefore, which is not consciousness for it is necessarily moment to moment only and not continuous, because it is related to the real, the dynamic and ever-changing, life or the living, and not to the unreal, that is, the ideated, the conceptual or the imagined. It is merely ideas, concepts, and products of imagination which make up the content of "the world's religious traditions." All of their rituals, beliefs, dogmas, and practices are merely mind-made, hence related to awareness which is consciousness and not to awareness which is not consciousness.

3.7
Chopra: Even if I see and hear nothing, I am still myself, an eternal presence of awareness .... A person who knows himself as spirit never loses sight of the experiencer in the midst of experience.

Commentator: Given what Chopra often says, one can say that "I" and "person" in these statements are the so-called Self. It is a so-called higher self which is a supposed "eternal presence of awareness," but self is never actually or really related to the eternal or the living but rather is always a mere creation out of a partially functioning mind. That which is living and can be called spirit is not anything which can be known. What a person or self, something unreal, can know is only a projection out of thought, out of a partially functioning mind. When there is awareness which is not consciousness, and not mere knowing by a knowing self on the level of awareness which is consciousness, one's mind is beyond the dichotomy of or division between the observer and the observed, the experiencer and the experienced, the knower and the known. Only when the experiencer is the experienced and not situated apart from what is being experienced is there the full and deep understanding which is vouchsafed only to a truly meditative mind.

3.8
Chopra: A key lesson in ... J. Krishnamurti was this: "Time is the psychological enemy of man," meaning that we are psychologically undermined and deprived of our real selves by the feeling that time is an absolute over which we have no control .... It is possible to have actual experiences of timelessness, and when that happens, there is a shift from time-bound awareness to timeless awareness.
Commentator: Here there is serious misunderstanding of something said by Krishnamurti. When Krishnamurti said that time is the psychological enemy of man, he was not at all suggesting that there are "real selves." Time is indeed not the absolute, but what often wreaks havoc in the lives of people is a lack of understanding of a proper relationship of mind to time and the timeless, is confusion of time and the timeless. Confusion is there when one talks about "experiences of timelessness." The timeless is not what can be experienced, just as the unknown is not that which can be known. To talk about "a shift from time-bound awareness to timeless awareness" is to indicate that one thinks movement from the known to the unknown is possible. Such is not the case. The known, time, must cease for the unknown or the timeless to be.

3.9
Chopra: The most creative people in any field ... grow with full consciousness that they are the source of their own power .... They self-refer: They place the highest trust in their own consciousness.
Commentator: Real creativity is related to a meditative mind, to profound intuition and deep understanding on the level of such a mind. On this level of the mind there is no self to grow, to enhance, to aggrandize itself. What a partially functioning mind might call full consciousness is not at all the profound awareness of mind beyond all consciousness. Self-referrers are merely selves who live, not out of deep awareness and richness of living, but merely superficially, in relation only to a partially functioning mind and to all of the superficialities related to such a mind, including all of the illusions which it can and does create.

3.10
Chopra: To want as much life, creativity, and wisdom as possible is very desirable .... Active mastery means having autonomy over one's life and circumstances, not power over others.
Commentator: Fullness of living, creative action, different from activity which is called or considered to be creative, and profound wisdom are of

mind in silent communion with reality or the living in all its depths, variety, and richness. If there is wanting or desiring, any "active mastery" of something, self is there, and, so long as self is there, there cannot be profound living and deepest wisdom, there cannot be really creative living. Merely to exercise autonomy, whether related to having power over others or not, rather than dependency as regards life and circumstances, amounts to a preference for one of two opposites, and such does not have anything to do with living richly and with profundity of awareness, understanding, and wisdom.

3.11
Chopra: While we are still unrealized, life is a struggle ... the technique that finally works [is] learning to accept your life not as a series of random events but as a path of awakening whose purpose is maximum joy and fulfillment.
Commentator: There is no point in talking about realizing ourselves for our so-called selves are not real things. To strive for such realization or fulfillment is to engage the mind in useless struggle, is to make life a struggle. Rather than seeking and striving for such merely mind-projected realization, one can choicelessly and observerlessly look at and be fully attentive to life, the living, and so understand it. Such understanding is in and of itself the meaning of life. No technique brings such understanding, such meaning. It is only a partially functioning mind which accepts or rejects, and any path it follows to so-called awakening is illusory. Joy is not something which can be attained by any partial functioning of the mind. It comes to the mind when unsought, when there is no effort and struggle to attain it, when, this is to say, there is no partially functioning mind or self which strives for so-called fulfillment of itself.

3.12
Chopra: You can learn to take your awareness into the region of timelessness at will – meditation is the classic technique for mastering this experience.
Commentator: There can be real timelessness only when there is no self to learn in relationship to the conscious awareness which is of self, of the known, of time. True meditation is possible only when there is no meditator, no self, when there is no experience to condition the mind, no matter what the source of the experience be – self, others, culture, environment.

3.13
Chopra: The quantum worldview is not a spiritual one in its equations and postulates, but Einstein and his colleagues were united by a mystical reverence for their discoveries.
Commentator: A statement like this reflects a failure to realize that there is dualism of time and the timeless, of time and eternity. Because of this dualism, one can see the difference between a mind which functions partially and a mind which is meditative. If one does not see this difference, one can easily confuse time and the timeless, and mind which functions with mind that is silent. Such confusion is there whenever there is an attempt like this one, namely, to integrate science and mysticism. Both science and mysticism are creations of a partially functioning mind, but, while the first has worth or value, albeit of a relative kind, the latter is worthless, of no value at all since it amounts to the futile attempt to go from the known to the unknown, from time and the relative to mystery, that is, the unknown, the timeless, the absolute. There is mystery, the unknown, the absolute only when there is an end to time, to the known, to the relative.

3.14
Chopra: A theory of everything is considered a viable goal by such notable thinkers as Stephen Hawking and Roger Penrose, [and] the spiritual masters ... realize that the unified field exists in, around, and through them.
Commentator: Once again here there is confusion related to a failure properly to see the difference between time and the timeless and between mind which functions partially and mind which is silent. Theory is appropriate as regards science, the known, time, the relative, but not in relationship to mystery, the unknown, the living, the timeless, the absolute. Were there ever to be "a theory of everything," this would have to do only with time, the relative, the known, and not with mystery, the living whole, call this spirit if you wish.

3.15
Chopra: Time seems to flow and move; clocks tick off their seconds, minutes, and hours; aeons of history unfold and disappear. But ultimately, Einstein held, this vast activity is all relative, meaning that it has no absolute value.
Commentator: Having said this, one can go on to say that beyond time, beyond history and its aeons, beyond all activity which is indeed all

relative and devoid of ultimate meaning is the timeless, the absolute, the living whole, with which only the quiet and not the partially functioning mind can commune. When there is such communion of the truly meditative mind, possible only when there is an end to time, history, activity, the relative, in a word, the lifeless and thus the dead, then there is the timeless and the action which is love, then there is the absolute, that is, the living.

3.16
Chopra: The ancient sages believed that direct knowledge of timeless reality is possible, [and] Einstein himself experienced episodes of complete liberation from space-time boundaries.

Commentator: The ancients wrongly equated all awareness to conscious awareness, all of the mind to its functioning parts, and so, not surprisingly, they spoke about "knowledge of timeless reality." There cannot be knowledge of timeless reality, however, since knowledge, at least in its usual or ordinary sense, is limited to time, the relative, experience, the known. There can, nevertheless, be awareness which is not consciousness of such reality when mind is in a truly meditative state. If, in the case of Einstein, there were indeed "complete liberation from space-time boundaries," this was not in relationship to experience, the dead and gone, the lifeless. Also, if there were such liberation, this was when Einstein was not doing science but rather when his was a state of mind transcendent to that which is there when science is being done.

3.17
Chopra: Unbounded intelligence, freedom, and power are inherent in the unified field that Einstein and the ancient sages shared in their vision.

Commentator: Any envisioned field, unified or not, is not that which is real but rather what is mere projection from out of a partially functioning mind. Deep intelligence and total freedom of the mind are there only when there is mind which is meditative and not when it is partially functioning. To attempt to bring Einstein as scientist and ancient sages together is to fail to see the difference between mind which functions in relationship to time, the known, as it does when one does science, and mind which is meditative and related, not to accumulated, so-called wisdom of the ancients, but to wisdom which is there moment to moment and so not at all continuous. It cannot be continuous because it has to do, not with what is dead, inert, abstract, and so on, but rather with what is living, hence that which is always dynamic and ever-changing.

In the passages quoted from Chopra, then, we see his constant equating of all awareness to conscious awareness. We also see that all of the mind is understood by him to be a partially functioning mind. He always assumes the presence of a "real Self," and thus, if he were asked if self, that is, a partially functioning mind can be or become a truly religious mind, his answer would be that it indeed can. In his book, we also see that he is one of those contemporary writers who believe that science, when wedded to mysticism, can and eventually will fathom the mystery which is life, that it can and eventually will lead to full and complete understanding of the eternal. Both of these answers, it has already been suggested, are illusory. The same or similar illusory answers to our two questions, we shall see, are given in the writings of Ken Wilber.

3.18
Wilber: Rationality (and vision-logic) might somehow be a *necessary* prerequisite for higher or transpersonal development. [Vision-logic] is integrating reason, [and it is needed for] a global spirituality.
Commentator: Because of the one real dualism, namely, the temporal and the eternal, it is important to see the difference between a partially functioning mind and one which is really meditative. The former relates to time and the latter to eternity. If one unduly mixes up and confuses the two, then there is illusion and lack of understanding, whether intellectual or of a meditative mind. So-called vision logic is an example of such confusion and illusion. Vision is always related to what is unreal, but in Wilber it is supposed to be really related to the transcendent, the eternal. Reason properly relates only to the temporal, but in Wilber, when it is aligned with vision, it is wrongly related to the transcendent, the eternal. Related to this confusion of the unreal with the real, is the confusion in Wilber which leads to his talking of the personal as something related to a certain level of consciousness and of the so-called transpersonal as related to a higher or deeper level of consciousness. Actually, this confusion is simply another example of the same thing, namely, unduly trying to join the temporal and the eternal. Reason is limited, and, whether joined to so-called vision or not, it cannot effect the integration which is possible only when mind is in a meditative state. Such a state is there only when reason and all of its fashionings come to an end. Any so-called spirituality, called global or not, is related to the illusion that there is a

Psychology: Science or Meditation? 99

real, higher or deeper self, properly called spirit. It has to do, not with what properly relates to the eternal, namely, a truly meditative mind, but rather it has to do with what relates merely to the temporal, that is, a partially functioning mind. In Wilber, so often, the partially functioning mind is functioning inappropriately, and this whenever there is disregard for its proper limits. Functioning of the mind which relates to belief in a real self, called spirit, is improper and inappropriate.

3.19
Wilber: Models of development that acknowledge higher or deeper domains [of consciousness] than the typical egoic state ... we must ... aggressively defend.
Commentator: It is a partially functioning mind which develops models. The so-called higher or deeper domains of consciousness which such a mind either does or does not acknowledge, which either does or does not defend, are not at all the depths of reality with which only a truly meditative mind communes. Self in some form is necessarily there when there is any state of consciousness. In any such state self in some guise is there to be aware. It is pure illusion to think that the self which is there when there is any so-called deeper, higher, or more profound domain or state of consciousness is other than the self which is there when there is the so-called "typical egoic state" of consciousness.

3.20
Wilber: The absolute ... is ... disclosed in contemplative awareness invoked by ... meditation, but what is disclosed cannot ... be further stated.
Commentator: The absolute cannot be "disclosed in contemplative awareness" for such awareness is consciousness, awareness, therefore, limited to the partially functioning mind and the temporal and not at all related to a truly meditative mind and the eternal. The absolute is the living, the unknown, real mystery, the really new, and thus it does not in any way relate to what is dead, the known, the old, the experienced, all of those things, this is to say, which are the focus when there is contemplation, one of the many possible activities of a partially functioning mind.

3.21
Wilber: The new canon ... is perhaps the only arch-view with a genuine chance of integrating science and spirituality (modern science is wedded to an evolutionary worldview; the Great Chain of Being, if temporalized, is quite consistent with that evolutionary view).
Commentator: "Canon" and "view," whether arch or not, relate to a partially functioning mind. Such a mind cannot at all effect integration as regards living, relating. To attempt to integrate what successfully and appropriately has to do with the temporal and spatial, namely, science, with so-called spirituality lands one into illusion since any such attempt is an instance of the known trying to know the unknown. Such an attempt is simply related to a projection of a partially functioning mind striving for the impossible, that is, striving to encapsulate the eternal, the whole, in time, in constructs which are necessarily related only to time. Sometimes constructs of a partially functioning mind are of value, have a necessary usefulness, but in a case like this one, they have no value at all, hence are completely useless or superfluous.

What Wilber has coined "vision-logic,"[6] said by him to be necessary for a new, all-encompassing spirituality is just such a projection out of a partially functioning mind as its strives to attain to the whole. Futile is any attempt of an abstract projection of the mind trying to grasp what is living, what is whole.

3.22
Wilber: The new canon alone ... can tie all ... great [spiritual] traditions ... into a coherent and persuasive holism ... can integrate ... subjective ... intersubjective ... and ... objective ... accounts of consciousness .... [This] canon translates directly into a genuinely integral spiritual practice, covering self and culture and nature ... uniting the best of premodern wisdom with the brightest of modern knowledge.
Commentator: "Canon" relates merely to a partially functioning mind, mind which relates, and can only relate to experience, time, knowledge, awareness which is consciousness. Real "holism," integration on the level of a truly meditative mind, is not a fusing, an integrating process effected by a partially functioning mind. Rather, such "holism" is there only when there is real transcendence beyond opposites like the subjective and the objective and like self and culture, that is, only when these opposites are negated by a silent mind seeing them for what they are, seeing that they are not real. Such transcendence is beyond all effort and activity of a partially functioning mind, hence beyond all practice and discipline. Only when

Psychology: Science or Meditation? 101

there is such transcendence is there wholeness which is not mere projection out of a partially functioning mind effected by a combining of so-called "premodern wisdom" and so-called "brightest of modern knowledge." Futile are all attempts to integrate so-called spirituality or so-called ancient wisdom with science. Because of the real dualism of time and eternity, one must not mix science with religion, with any so-called spirituality. Any such mixing would be the confusion of what is of value, although limited and time-related, namely science, with what is of no value at all, namely, so-called "premodern wisdom." So-called wisdom is not valuable but rather futile and illusory since it is an instance of what is merely time-related attempting to reach or capture what is eternal. Only when this and similar attempts are ended can there be a truly religious mind, a mind which can function partially when such functioning is of value and necessary but which is totally silent or completely meditative when it need not function partially.

3.23
Wilber: The traditional realization of enlightenment [is] a recognition of spirit, by spirit, as spirit.
Commentator: Reality characterized as spirit is not reality but merely what a superficially functioning mind projects to be reality. When mind recognizes anything, it is merely finding something in relationship to the known, the experienced, the dead past, hence that which is not at all real. The so-called spirit which recognizes is nothing else but the self, a mere construct of thought and not anything real.

3.24
Wilber: The spiritual line moves from ... *prepersonal* ... to ... *personal* ... to ... *transpersonal* ... runs through subconscious to conscious to superconscious, [from] preconventional [to] conventional [to] postconventional, [from] *egocentric* [to] *sociocentric* [to] *worldcentric* .... The worldcentric ... is the platform and the gateway to higher spiritual domains.
Commentator: Any and every movement along a so-called "spiritual line" is related to the illusory and has no relationship at all to living truth. What would be said to be significant points on such a line of development are indicated by what are called the prepersonal, personal, and transpersonal, and by what are alternatively called the preconventional, conventional, and postconventional, but all of these points, as much as the line itself, are merely projections out of a partially functioning mind. A

mere making of divisions within the whole of consciousness so that there might be talk about levels or degrees of consciousness means one is still locked within awareness which is consciousness, with no seeming realization that there can be awareness which is not, which is beyond, consciousness. Whether one is supposedly at the egocentric, sociocentric, or worldcentric point within consciousness, ego or self is always there, and, when self is there, there cannot be real love. Only when self and all which that self projects is ended by awareness which is not self-related can there be that which is higher or deeper, and this higher or deeper is not any so-called domain fashioned by a partially functioning mind.

3.25
Wilber: Your union with Ground ... is always present, but it can be either conscious or unconscious.
Commentator: So-called conscious awareness with the so-called Ground relates to mind which projects a "you" out of itself, that is, a self said to be aware, and which, as a mere projection out of a partially functioning mind, is not anything real. Also a projection out of itself is any union of which this self is said to be aware and thinks it enjoys. Such a so-called union is not at all truly silent mind communion with the whole, the Ground which is real, the living.

3.26
Wilber: Growth [of the self] will involve ... expansion into wider and deeper and higher modes [of consciousness] ... development [of the self] from narrower to wider, from shallower to deeper ... from matter to body to mind to soul to spirit.
Commentator: So-called growth of the self is growth of something which is not real, something which is a mere creation of thought, and so expansion or growth of consciousness, of this unreal self has nothing to do with what is living, that which is deep or profound, that which is mystery and hence the unknown, the new, the ever fresh, the dynamic and ever-changing. Movement from one so-called level or stage of consciousness to another is not movement to that which is profound or deep, namely, the eternally alive, the living, but merely at best from the shallow to the shallow, from one mind-created fragment, namely, matter or body to another mind-created fragment, soul or spirit. Though there is one aspect of the one psychophysical organism which is man that can be called 'body,' and 'mind' the other, soul or spirit is something unreal, that is, a mere projection from out of thought, out of a partially functioning mind.

3.27
Wilber: Rest in Emptiness, embrace all Form.
Commentator: Self rests and self embraces, and so this Emptiness is something unreal, as unreal as any forms which self fashions. Indeed, self – higher or lower – is merely one of the many forms fashioned by thought, by a mind which partially functions. It is not when there is embracing of something, but rather when there is a negating of everything which is partially functioning mind fashioned, of each and every form which a partially functioning mind creates, that the immeasurable, the absolute comes into being.

3.28
Wilber: In the advanced stages of spiritual practice, the self will ... recognize the changeless ... its own primordial nature, the ever present Emptiness, [and when there is such recognition] God-consciousness has become ... a timeless trait.
Commentator: The changeless is that which is inert or dead, and that which is dynamic and ever-changing is the alive, the living. What is really alive, however, is not a merely mind-projected something called the alive or the living, and it is never that which can be recognized since it is ever new, ever fresh, ever timeless. Neither is it something to be realized as a result of practice of some kind. Consciousness cannot lead to awareness of the timeless. Time and consciousness must be ended if the timeless is to be.

3.29
Wilber: So many great spiritual philosophers engaged in ... incredibly *intense* polemic.
Commentator: To make this kind of statement and implying that the engagement referred to is a necessary and valuable thing is to show that one fails properly to distinguish between or improperly attempts to relate time and the timeless or the eternal. It is also to show that one fails properly to see the difference between a partially functioning and a truly meditative mind.

If so-called enlightened awareness has a "wrathful aspect"[7] opposite to its benign aspect, then such awareness is awareness which is consciousness, and it relates to a partially functioning mind. Any such so-called enlightenment is not at all related to a truly meditative mind to which alone profound enlightenment comes. This profound, real enlightenment is not at all to be confused with the level of mind on which there is discrimination and the making of judgments, with that level of the mind to which morality in the ordinary or usual sense relates.[8]

Both those who display and those who are said to lack the so-called wisdom and strength of moral character which are totally related to a partially functioning mind, but not at all to a meditative one, are not really and deeply wise and moral if and when they are wrathful and fierce in their judgments.[9] Only when there is no wrath, no fierceness, nor their opposites, is there a truly silent, meditative mind.

If a green, lush, and fruitful "tree" is a good symbol of that which is living,[10] then it is not an appropriate symbol for the movement called transpersonal psychology for the transpersonal, as much as the personal, is not anything real but rather something which is mere projection out of a partially functioning mind, mind functioning inappropriately, however, and thus relating itself to that which is dead. A piece of dead wood could be said to symbolize this movement since its focus is on the past – experience, the known, mind fragments.

3.30
Wilber: People too often imagine that "choiceless awareness" means making "no judgments" at all. But that itself is a judging activity. Rather, "choiceless awareness" means that both judging and no judging are allowed to arise, appropriate to circumstances.
Commentator: Only one who would say that all awareness is consciousness and all of the mind is merely a partially functioning one would make a statement like this. This statement suggests that someone like Krishnamurti who spoke about choiceless awareness in relationship to a truly meditative mind was mistaken when he described such awareness as he did. To say that one who says that choiceless awareness means making no judgments at all is simply in the realm of imagination is to indicate that one does not realize that besides the partially functioning mind which can judge and imagine, and so on, there is the totally quiet, silent mind, mind which, though it never judges, understands fully, is profoundly wise, and really loves.

3.31
Wilber: To manifest a genuine spiritual consciousness ... men need to transform their agentic selves ... women ... their permeable selves (from selfish to care to universal care to spiritual recognition).
Commentator: Consciousness at any level does not, indeed cannot, relate to what is real, what is genuine. Engagement in activity which

supposedly will transform self, whether so-called and supposedly agentic or permeable, is futile and illusory.

3.32
Wilber: Female [spiritual] practices ... involve ... *transformative* Agape [and] offer a stunning contrast to the more traditionally ... agentic, and Eros-driven modes of spirituality typical of males.

Commentator: Real love is related to the whole, or rather is the whole. To speak of feminine Agape as different from masculine Eros is to speak of supposed types of love. Types of love, however, could not be anything more than parts, and parts are not real. Real love is not what is attained by practices, by spiritual exercises, by effort, by discipline. Real love is revolutionary, not merely reformatory. Nothing which a partially functioning mind does is profound. Only revolution in the mind and not mere mind transformation is profound, and to shift from one world-view to another,[11] that is, from one partially functioning mind produced system of constructs to another is not at all related to the richness and depths of reality, and not to the profound depths of mind which is totally silent.

So-called "female practices" on the part of a so-called feminine, permeable, connected self which is said to commune rather than being active like the so-called masculine agentic self are as much related to the unreal, that is, to mere formulations or projections of a partially functioning mind as are so-called "agentic and Eros-driven modes of spirituality." Maleness, just like femaleness, is something rooted in physical reality, but masculine behaviours, like feminine behaviours, are merely related to conditioning of the mind, to things which are mere arbitrary creations of a partially functioning mind, like "modes of spirituality." Wilber is wrong when he talks about such mere creations or fashionings of a partially functioning mind as if they are realities. The oppositions which we read about in the writings of transpersonal psychologists like Wilber, for example, agentic and permeable selves, Agape and the Freudian Eros, and so on, relate not at all to reality, to what *is*, but rather merely to what is fashioned on the level of a partially functioning mind.

3.33
Wilber: The permeable, [connected] self ... undergoes growth, development, and transformation. [Although both the permeable and the agentic selves grow and develop], males [tend] toward agency and Eros, females toward communion and Agape.

Commentator: Self in any form is unreal. Hence, any so-called growth, development, and transformation of self is related to a process which can only lead to nothingness, to illusion. Masculine and feminine are mere creations of a partially functioning mind and so too are the so-called permeable and agentic selves. To distinguish between these two types of selves and between Eros and Agape is to distinguish between mere images, things unreal. Images are not reality, and these particular ones relate to the setting up of differences which are not real and which, when taken to be real, cause hurt, harm, confusion, and conflict. They do not at all relate to the living, loving human being nor to real love which does not have parts and which does not divide itself up into types.

3.34
Wilber: Extraordinary women mystics say [that women] have a relational, embodied, permeable self, [and they talk about transforming] that self and [rendering] it fully transparent to the Depths of the Divine.
Commentator: Self is not anything real, and this whether it is the so-called feminine self which is said to be relational, embodied, and permeable or the so-called masculine, agentic self which is opposite to the feminine. Self transformed is still self, and, where there is self in any form, be it the one which is there before so-called transformation as well as the one there after such so-called transformation, there cannot be anything related to what is deep and profound but only that which is superficial, and this because it is always and necessarily unreal, merely mind-made, mind-projected.

Before returning once again to our two questions, we look finally in this chapter at parts of *Crazy Wisdom,* mentioned at the beginning of this chapter. Nisker's is a particularly unique way of expressing a prizing of right rather than left-brain thinking. However, as he speaks about life and its meaning, his is a failure to make a fundamental difference between time and the timeless and between a partially and a totally functioning, truly meditative mind. Following are some selected statements from *Crazy Wisdom* along with commentary on them to show that this is indeed the case.

3.35
Nisker: The artistic side of crazy wisdom, like its spiritual counterpart, consists primarily of personal visions.

Commentator: Visions relate to the unreal. There cannot be living truth, truth about the living, mystery, the unknown so long as there is merely a partially functioning mind, one aspect of which is right hemisphere thinking in terms of the images, symbols, and visions which are either artistic or related to a so-called spirituality which is either traditional or so-called New Age or contemporary.

3.36
Nisker: Crazy wisdom is intrinsically contrary .... Often, followers of crazy wisdom challenge the established order .... At other times, they walk away into the mountains to live a simple life.
Commentator: What is contrary, intrinsically or not, is related to a partially functioning mind, mind functioning rationally or irrationally. A mind which is contrary is a mind in a dualistic state of awareness, a mind merely on the level where opposites are created, for example, any established order as opposite to what might replace it when it becomes disestablished, and which also, like any established order, is merely something fashioned by a partially functioning mind. Simply to retreat to the mountains or to the ashram is mere reaction on the part of a mind which operates in terms of some fragment, like will or imagination, is engagement in what might be called a personal search for meaning, is perhaps the result of some decision or so-called personal choice not only not to be part of this world, that is, the established order, but also not even to be in it. It is only the truly meditative mind which does not react, does not oppose, but which, while in the world but not of it, understands fully and lives profoundly.

3.37
Nisker: Crazy wisdom is about another way of knowing. Some call it intuition or vision .... Some modern seekers say crazy wisdom comes to us through a different part of the brain, perhaps the mysterious right hemisphere.
Commentator: Given his description of those who create and present crazy wisdom, Nisker is wrong to include Krishnamurti as one of these, to refer to him as one of those he calls "contemporary holy fools."[12] Krishnamurti was not one of those who merely seek another way of knowing, knowing in relationship to right rather than left-brain partial mind functioning. The level of the mind that Krishnamurti was always on when the matter was living was that of meditation, and, when there is real meditation, there is no vision and nothing of the visionary but rather a

direct facing of what *is*. Only when there is such facing do fullness of understanding and profound wisdom come to the mind.

3.38
Nisker: We need to ... shut off or turn down the grinding noise of the rational, analytic gears ... through prayer, meditation, nature, art, poetry, stories, and song.
Commentator: Self or a partially functioning mind is still functioning partially when there is such shutting off or turning down. To move from concepts to images and symbols, from the rational activity of the left hemisphere of the partially functioning mind to right hemisphere functioning, that is, for example, praying, so-called meditating, singing, story-telling, and so on, is not to stop all partial functioning of the mind so that there can be the understanding of living truth, the loving intelligence, and the profound wisdom of the truly meditative mind. Only such a mind lives deeply rather than superficially, in communion with the really good, with real joy, peace, and order touching the mind and one's way of living in the world.

3.39
Nisker: The self we think we are is not the real self, or we have many selves .... Of course, there is always another side to the story.
Commentator: Here there is the false dualism of ego, a "self we think we are," and a supposedly "real self," a soul, the Atman. Whether there are few selves or many selves, self is still there, and, so long as self is there, one is necessarily on the level of mind where there can be one side opposed to another side or to other sides. Beyond all oppositions, all dualism except for that of time and the timeless, beyond all the limitations which are associated with a partially functioning mind is the truly meditative mind, are all the treasures, so to speak, which come to or are only given to a mind which is absolutely quiet, totally silent.

3.40
Nisker: Great artists stop the mind; their work brings a sudden, intuitive revelation that allows the "subconscious" or the "superconscious" to come to the forefront.
Commentator: If self – the artist or some other – attempts to stop the mind, there is no real stop, that is, quietening of the mind so that it is totally, absolutely silent, meditative. Rather, there is merely movement from one kind of partial functioning to another, perhaps from left to right

hemisphere functioning, or from right to left. A so-called intuition associated with any aspect of a partially functioning mind is not at all the integral intuition which is possible when there is the state of the truly meditative mind. Any so-called revelation which is said to come to a partially functioning mind is not of what is new and living, the unknown, real mystery, but rather it is related necessarily to what is already known, what has been experienced, what is recognized, in a phrase, to what is old and dead. Strictly speaking, therefore, only intuition from out of a really meditative and not out of a conscious, partially functioning mind is truly, really creative. What is truly creative relates not to awareness which is consciousness, not to any division within consciousness made by a partially functioning mind and called superconscious awareness, but only to awareness which is other than conscious awareness. Conscious awareness necessarily has to do with self there being aware, but, when there is awareness which is not consciousness, then there is no self at all there being aware.

3.41
Nisker: Techniques such as meditation ... might bring us to true sanity or consciousness, and a more integrated way of living.
Commentator: When there are techniques, related to so-called meditation or not, self or a partially functioning mind is necessarily there. If self is there, any meditation there might be would be meditation with meditator present. Any such meditation, since it would have to do with activity on the level of a partially functioning mind, would relate to what is superficial at best. Only meditation without a meditator can make for a fully "integrated way of living." Any so-called integration effected by self or a partially functioning mind is necessarily limited if not illusory. If "true sanity" relates to fullness of understanding and truly integrated living, then it is not merely related to consciousness but also and significantly to awareness which is other than consciousness.

3.42
Nisker: Meditators ... see through their individual obsessions and confusion to discover what is often referred to as our "true nature."
Commentator: A meditator, that is self, cannot discover anything new, anything which relates to what is living. It cannot see totally, with fullness of understanding, hence it cannot really see through the obsessions and confusion which are there in lives lived only in relationship to partial functioning of the mind. As regards living and relating, there is ever

conflict and confusion when there is only the partially functioning mind, whenever self, this is to say, is engaged in one or several of its activities. There is no "true nature," some supposed inner essence or soul, to be discovered by a limited, partially functioning mind. To think that there is relates to one example of the confusion which is there when there is an improperly partially functioning mind, when self is there projecting illusion out of itself.

3.43
Nisker: Meditation practice is an antidote to our desires and fears, a way to allow ourselves to be in the world as it is.
Commentator: Real meditation, that is, meditation related to the real, and not to some ideal, image, symbol, idea, or concept is not something which can be practiced. Whenever there is practice, there is a practitioner, that is, a self or a partially functioning mind creating conflict and confusion in living, in relating. An antidote is an opposite, and, when the matter is psychological, is living and relating, nothing related to living and relating can be fully understood and resolved so long as the mind is in a merely dualistic state, the state of mind wherein opposites reside. So long as the mind walks in a tunnel of opposites there cannot be living in the world as it is, in relationship to what is alive and living, to mystery, the new and the unknown but rather only living in illusion, in an unreal world, in a dead world.

3.44
Nisker: The theater of meditation requires that part of the mind be trained as an observer, known by various names: the "witness," "the other," or the "higher self."
Commentator: When there is a meditator engaged in so-called meditation, the meditator is indeed an observer, but an observer separate from the observed. This observer indeed is merely a witness witnessing the witnessed, a watcher separate from what is being watched, self in a particular guise perhaps called higher rather than lower. So long as the observed is separate from the observer, however, the witnessed separate from the witness, and so on, there cannot be the insight and understanding which are vouchsafed only to a truly meditative mind, a state of mind wherein there is no witness, no watcher, no observer, no meditator. Training of a partially functioning mind can never give the complete understanding and make for the holistic living which relate to the state of the truly silent, meditative mind.

3.45
Nisker: We know that all things require their opposites: dark is necessary to light, death is necessary to life .... To accept paradox is to accept what appears to be the fundamental law of nature and rule of life.
Commentator: What is true to say is that *physical* things can have opposites and that such opposition relates to physical laws of nature. This is not the case, however, as regards matters psychological. On the level of the living there are no real opposites. Thus, although physically speaking there is a separation between the life and the death of the organism which is man, psychologically speaking life and death are not separate. It is only a partially functioning mind which will separate them, cause division which is its mind-created opposition between them. Also, real living love has no opposite. Only a partially functioning mind will set up a so-called love opposite to hate. Realizing the truth of these matters is not a matter of acceptance of something but rather of totally seeing out of a really meditative mind and so then understanding the truth of what *is*.

3.46
Nisker: Science may have finally moved into the realm of myth or poetry .... Perhaps contemporary scientists are simply giving us new symbols for the incomprehensible mysteries of the universe.
Commentator: To say this suggests that science is moving away from exclusive left-brain thinking and the objectivity associated with such thinking. It might well be the case that, if and when science does this, and thinks that it is thereby coming closer to an understanding of what is living, to measuring the immeasurable, to fathoming the depths of living mystery, it fails properly to understand its true nature, that it is limited, that it can never fathom the richness and depths of what is living. In any case, any such move on the part of science could never result in the discovery of anything but "symbols" projected out of a partially functioning mind. It cannot relate to the discovery of living mystery but merely to a finding of answers to what is problematic as regards the physical world, what is time and space related. Of worth and deep significance is meditative mind communion with the eternal, the living mystery which necessarily is incomprehensible to a partially functioning mind, whether scientific or some other. Since the scientific mind is always and merely partially functioning, of course it can never comprehend the totality, the truly universal, the eternally living in all its depths and vastness. Such is possible only for mind in a truly meditative state. Such

comprehension is the meditative mind's communion with the absolute, mystery, the immeasurable, the ineffable.

3.47
Nisker: Many critics believe that by reducing everything to matter and energy, science has robbed us of spirit and mystery.
Commentator: It is only when there is a failure properly to make a difference between time and the timeless, and a failure properly to see science as necessarily limited to time and having nothing at all to do with the timeless, that one suggests that science is such a robber. Science might well be right to reduce all things *physical* to matter and energy, but, if it does this, this does not mean there must be denial or shunning of the mystery which is living. Communion with this mystery is there when, in addition to a mind functioning partially when such functioning is appropriate and necessary, there is also and otherwise a truly meditative mind, a completely silent, quiet mind. If, in this statement, so-called spirit is something opposed to matter, this is a false dualism and the mystery spoken about is not real mystery.

3.48
Nisker: The observer and the observed seem to be locked in an eternal dance with each other, both unable to stop long enough for a clear picture to be taken.
Commentator: Separation of the observer and the observed applies only to time and to mind functioning in relationship to time. When there is a truly meditative mind and communion of such a mind with the eternal, there is no observer separate from the observed. Then the mind is not locked in, not in any kind of prison created by a merely and always partially functioning mind. Total clarity of a fully functioning mind is there when there is the truly meditative mind in communion with the living, the absolute, the eternal.

3.49
Nisker: In the twentieth century the struggle was between the artist/humanist and the rationalist/materialist: two cultures, two minds.
Commentator: This struggle was not really between two minds but between two aspects or two forms of a merely partially functioning mind. What was not there, generally speaking, in the twentieth century as well as in the centuries which preceded it was the truly meditative mind. Never was there an understanding of that state of mind beyond or transcendent

to all forms or all types of the partially functioning mind. When there is this state of mind, there is no struggle of any kind at all because there is communion with the whole and not merely the parts of the mind which can struggle one against the other or others.

3.50
Nisker: In *Deep Ecology*, John Seed writes [that] the change that is required of us is ... a change in consciousness. Deep ecology is the search for a viable consciousness.
Commentator: A mere change in consciousness is not profound, revolutionary change. Really radical change is possible only when there is awareness or understanding that besides awareness which is consciousness there is awareness which is not conscious awareness. Besides a "viable consciousness," what is needed or what could be seen to be of immense significance and value to people is awareness related to the really eternal, the living, the new, the unknown, the absolute. Such awareness is other than awareness which is consciousness.

3.51
Nisker: There is no self; all we have to lose are our illusions. Death is simply a way of ending ... painful separation, a way of dissolving back into the flow, a homecoming.
Commentator: There is indeed no self which is anything real, but saying this must be understood as saying there is not only no ego, no lower self as anything real but neither is there any so-called real inner, deeper, or higher self, no soul, no Atman. This statement seems to imply a real inner self, there before the separation, forgotten but still there during the separation, and then integral once again after there is the dissolution of the ego, the lower self, and the homecoming, that is, a so-called coming back to that supposed place where there is indwelling of the so-called real, higher, deeper Self. When there is no separation between life and death, living and dying, there is no pain, no separation at all. Indeed then there is no time but only the timeless and so no process of dissolution and homecoming. There is no self to be dissolved, no self to come home. There is only mind which is truly meditative, which functions partially only when necessary. When there is no inappropriate partial functioning of the mind, there is only life eternal and no psychological pain and suffering which are time and partially functioning mind related.

3.52
Nisker: What we might call "consciousness" ... has not yet developed in us .... Another kind of intelligence might arise once we accept our ignorance and our limitations.

Commentator: Such a statement reflects a failure to realize that, besides consciousness and its necessary limitations, there is awareness which is not consciousness and to which the deep, loving intelligence of the really meditative mind relates. Such intelligence is neither left nor right-brain related but is of a level of the mind beyond all parts, all fragments.

3.53
Nisker: Accepting uncertainty as our philosophy might allow us to honor each other's stories more, delighting in all the bizarre and wondrous interpretations of the mystery.

Commentator: Rather than merely taking delight in right-brain, partially functioning mind created mere interpretations of so-called mystery, like those of Rumi, the Sufis, Zen Buddhists, and the likes of these, one can, in the state of the truly meditative mind, commune with real mystery. When there is real, direct communion, one lives in relationship to what is timeless, to deep wisdom, to fullness of understanding, to loving intelligence, to the new, the living, the absolute, the immeasurable and not to what is partial, fragmentary, limited, superficial, like any fashioning of a partially functioning mind, be that either left or right hemisphere functioning.

Such then are some writings which directly or indirectly relate to much of contemporary psychology. Ken Wilber, just like Deepak Chopra and Wes Nisker, would say, but perhaps not in these same words, that a partially functioning mind can become a truly religious mind. It can be on its way to becoming such a mind, Wilber would say, when there is vision-logic and subsequent movement to so-called psychic, subtle, and causal stages of consciousness development. Vison-logic, however, and what might come after it when there is movement from the formal-reflexive stage of consciousness to vision-logic and then to the psychic and even further beyond it, are, as has already been indicated, merely fashionings of a partially functioning mind, and, as such, they cannot really relate to what is the really living, the eternal, the mystery which is the unknown. Communion with the eternal, with real mystery is possible only when there is a really meditative mind.

Psychology: Science or Meditation?

As regards the second of our two questions, the writings of Wilber, like those of Chopra and Nisker, suggest that science, or at least the mind which creates science, when it is like the mind of the mystics or the ancient sages, is capable of and will eventually succeed in fathoming mystery, the unknown, ultimate reality. Such suggestion, however, relates to a failure properly to see the difference between time and eternity, between a partially and a totally functioning mind.

Given all that has been said in the commentary in this chapter, it is clear that the right answer to the question which is the title of this chapter is that psychology, at least as we see it in Chopra and in Wilber's so-called transpersonalism, is neither science nor meditation. Indeed, it is not at all related to either intellectual or living truth, and as such it has to do with what is unreal, what is false, what is illusory.

In the next and final chapter of this book, correct and appropriate answers to our two questions will again be highlighted as a final look at significant aspects of our contemporary world is taken. Out of this looking come not only appropriate answers to our two questions, but also appropriate answers to the questions about life and its meaning which people ask, but which, once asked, are usually either cynically dismissed as questions impossible to answer, or are answered with answers that are illusory and thus ultimately unsatisfactory. Full and complete answers to our two questions and to questions about the meaning of life or of living come only out of a truly meditative mind.

Chapter Four

Science and Silence

> Meditation is the purgation from the mind of all its accumulations; the purgation of the power to gather, to identify, to become; the purgation of self-growth, of self-fulfillment; meditation is the freeing of the mind from memory, from time.
>
> The state of the religious mind can be understood only when we begin to understand what beauty is; and the understanding of beauty must be approached through total aloneness .... Aloneness is ... not isolation .... To be completely alone implies that the mind is free of every kind of influence and is therefore uncontaminated by society; and it must be alone to understand what is religion – which is to find out for oneself whether there is something immortal, beyond time.
>
> J. Krishnamurti

An excellent overall portrayal of many major features of the contemporary world is given in "The Me Millennium" articles which appeared in *The New York Times Magazine* in October of 1999.[1] A reading of these articles from out of the state of the truly meditative mind, however, brings the realization that, though the description of the contemporary world given in these articles might be said to be objectively accurate and insightful, that, though it is indeed the case that ours is a world becoming more and more fragmented, divisive, and isolating, as these articles suggest, nothing is said which relates to the deep understanding and meaning which are possible when there is awareness on the level of the really meditative mind. It is this understanding and this meaning which alone can address the pessimism, cynicism, skepticism, and nihilism which today cast their spell all around in this world of ours. Such an address is profound and not the merely superficial reaction to pessimism and nihilism which one sees in the cautious, sober, but ultimately futile optimism and hope related to traditional religious stances or positions like that of John Paul II and the Buddhist Promoting Foundation.

Insightful description of the contemporary world is given when Andrew Cherlin talks about "wariness of organized religion and lack of interpersonal trust,"[2] when David Samuels speaks about young people seemingly without any "connection to anything larger than their own ... aims and preoccupations" and about people themselves as "the most profound of dissatisfactions,"[3] and when Camille Sweeney writes about "teen-agers" who "can discover themselves (or the many parts of themselves) by roaming the boards and the chat rooms, connecting, disconnecting, shooting questions out into the universe – and maybe, just maybe, receiving answers."[4] Not at all can one say that any of these people, who it might be said are representative of most, if not all of us, are aware of that deep level of the mind which relates to understanding beyond trust and distrust, beyond aims and occupations, and beyond any of the superficial answers which are possible when there is merely partial functioning of the mind and never that silence which is communion with the eternal and things which are of the treasure house of the eternal.

Mind which is merely on the level of opposites, specifically the individual as opposed to society, is indicated when the editor of *The New York Times Magazine* says "how hard it is, in a time of gathering global conformity, to find one's own way."[5] Ultimately, it is fragmentary, even futile effort which would merely "eschew the use of I" in favor of "*yours truly*," or would "replace the 'perpendicular pronoun' with the horizontal pronoun, *myself*,"[6] and this because to do so is merely to stay on the level of a partially functioning mind, and not to go beyond the level of opposites to that deep level of the mind which alone can commune with living truth, eternal and profound as it is. Staying always and only on the level of opposites is taken to be that which alone is possible when "at the end of the millennium, despite the collectivist counterreaction in the 20th century, it is still the individual who inspires the masses."[7]

It is mere movement from one pole in a set of opposites to the other which is described when Richard Russo indicates that people have expended a great deal of energy changing the "medieval mind-set" in relation to which "the individual" had very little "importance except as a ... cog in the great wheel that moved" in accord with the will of God.[8] Indeed, as Russo rightly points out, "the 'Me Decade' of the 1970's was little more than a particularly virulent manifestation of a 600-year philosophical trend,"[9] namely, movement away from emphasis placed on the object known, to emphasis placed on the subject, that is, the self, that knows. When there is self that knows facing object that is known, this is clearly indication that one is on the level of the dualistic mind.

What is called for now, it could be suggested, but nowhere indicated in the "The Me Millennium" articles, is movement beyond emphasis on either object or subject, or beyond any partially functioning mind attempt to bring subject and object together, to synthesize or amalgamate the two. Such movement is to awareness on the level of mind beyond all parts, all fragments, hence beyond object known and subject which knows, beyond observer separate from that which is observed, namely, the level of mind which is truly meditative, mind whose intelligence is living and loving rather than merely abstract and rational, mind which in its total wisdom allows for proper partial, rational functioning of the mind but sees when all such functioning is inappropriate, mind which is completely free because totally unconditioned, mind which is perfectly ordered, deeply and completely tranquil, hence loving and peaceful.

When there is true silence of the mind, one neither laments, as John Paul II, for example, might, the loss of a so-called but merely mind created objective moral order, nor does one strive for a humanistic "compensation" with respect to this loss, like that of Matthew Arnold.[10] Neither does one think there is great profundity of mind when Whitney Houston eulogizes love of self.[11] The truly meditative mind realizes that the self is ultimately nothing, and, if the self is ultimately nothing, then what nobility or wisdom can there be in such so-called love for the nothing, that is, the no-real thing which is something merely mind-fashioned, a mere image and not something real? This no-real thing which is the self is not the nothing, however, the no-thingness which is ultimate reality, that which is living eternally and profusely, life in all its abundance and richness, and in relationship to which there can be real love on the part of a truly quiet, silent, meditative mind.

Even though it can be said that the Renaissance gave rise to "individual self-consciousness,"[12] to "the flowering of the Self,"[13] so that now we have "personal identification,"[14] as well as last names, vacillation between individualism and collectivism characterizes the whole history of the Western World, indeed, the history of the whole world. Such vacillation is always there too in this our contemporary world. As Richard Russo says, "conceit of 'the self'" is neither "recent" nor "exclusively Western" for even "ancient peoples had a sense of self-consciousness."[15] Simply to continue to move around one or other of the two poles which are individualism and the collectivity, however, or to move from one to the other in effort to attempt to find or to make sense and meaning of living or relating is to engage in activity which cannot give profound understanding and insight and cannot bring about the living of a deeply meaningful life. What such movement might at best do, on the one hand, is "yield" so-called "metaphorical and interior meanings,"[16] that is, subjective or so-called personal meanings, or on the other hand, some so-called objective meaning, perhaps related to the Christian faith and philosophy of being of someone like John Paul II. All such so-called subjective or objective meanings are not really profound and significant for they relate to fragmentary or partial functioning of the mind and not to mind in a meditative state.

That solitude in which "introspection and art (and vice)" can "flourish"[17] is not the aloneness of the truly meditative mind. Any so-called "cavelike retreat that insures the vast inner open field of privacy"[18] as well as the so-called soul which desires to or in fact does escape to it are merely fashionings of an imaginative or a purely speculative, partially functioning mind. Such an escape is from what *is*, hence is movement to what is false, illusory, unreal. The final outcome of such movement might be some seeming consolation for the individual as opposed to the collectivity. What this movement cannot at all lead to, however, is the deep and ultimate meaning which are vouchsafed only to a really meditative mind. In "the cave, the belly, the refuge, the retreat"[19] which relates to a so-called reclaiming of one's Self, Self can indeed exercise an "atavistic enjoyment"[20] of things it has gathered about itself, but what one must realize is that Self is not anything real. It is unreal, that is a mere creation out of thought, something merely fashioned by a partially functioning mind. In the solitude of the cave, the belly, the refuge, the retreat to which the Self, not anything real, escapes there cannot be anything new but merely that which is of the dead since it is necessarily of the known, the measured, the experienced. Whatever flourishes there will be the false, the illusory, and not what is living, of the eternally now moment, of the unknown, the immeasurable.

The unreality but nevertheless merely practical, partially functioning mind usefulness of "personality" and "selfhood," that is of the self, is indicated when one realizes, as Mihm and Brown have written, that "at one time personal identification was a novel, even revolutionary accessory of selfhood. For much of the millennium, ordinary people did without the trappings of personal identification."[21] When there is true revolution in the mind, there is no "cult of personality,"[22] no "Byronic self-aggrandizement" as a "path to immortality."[23] Then there is no self with supposed psychological or spiritual reality at all but merely and only self as related to necessary partial functioning of the mind, functioning related, for example, to the physical, practical aspects of living.

One who is aware on the level of the meditative, really religious mind does not prize "personal growth"[24] and does not at all "thrill at the notion of... donning a new, improved identity ... to make" the "'self' more worthy of notice and admiration"[25] for it deeply realizes the ultimate nothingness of the self and looks to the eternal values rather than to sensate ones related to self as it truly lives a life which is profoundly understanding, loving, wise, peaceful, and ordered. Only a mind which is not really meditative thinks that one must choose between "esteem" for one's "self" and "floating face down in the pool."[26] A truly meditative mind is beyond the confusion and lack of understanding which are there when mind thinks it must choose. Equally confused and in conflict is a mind which entertains "intimations of immortality" because it thinks that "the self" which is "changeable and improvable" is "in some sense immortal."[27]

Symptomatic of the falseness of self-consciousness dominating the contemporary world is the penchant for "personality tests"[28] on the part of institutions and hiring agencies. That the self is not anything real, and thus that self-consciousness does not relate to anything real but rather only to what is merely mind-fashioned is indicated when Margaret Talbot points out that "the history of personality testing neatly reflects ... shifting notions of the optimal self."[29] Self is indeed merely a notion and not a reality.

The falseness in much of the world which is ours relates to the kind of thinking and activity exemplified in the fact that, though there are "biases in personality tests" and they do reflect elitist "values and experiences," such "testing" will go on since "institutions" are "hooked on" them and they promise "teasingly ... to satisfy" people's so-called "deep human curiosity."[30] How far we are from loving intelligence and the fullness of truth and beauty which relate to such intelligence!

Russo writes: "So here we are ... at the end of the millennium, having lived with the consequences of a me-centered universe long enough to wonder whether this direction we have been heading in for so long is a good thing."[31] Although we have lived this way, and in opposition to the first millennium's emphasis on object, the group, the commune, the collectivity, a truly meditative, really religious mind realizes that there is a totally radical, new way of living beyond the two different opposed emphases of the last two millennia. The profound question of such a mind is not whether either one or the other of the two emphases is good or bad, is not whether either one is better than the other, but whether or not there is a way of living related to a non-dualistic state of mind beyond dualistic thinking in terms of the good and the bad, the one and the other, the better and the worse, and so on. Such a mind is aware that there is the eternal beyond the temporal, the immeasurable beyond the measurable, the unknown beyond the known, ultimate good beyond good and evil, and silence beyond science rather than simply being conscious and knowing in relation to the subject – that is, the "me," the self – facing an object. This mind, furthermore, is aware that what is truly significant is living in relationship to deep understanding, to the profound wisdom and insight which are only of a silent, meditative mind, of a truly religious mind, and not of a mind which functions partially or fragmentarily in relationship to what is ordinary, organized religion or in relation to some form of partially functioning mind opposition against or indifference to it.

Only the "eyes"[32] which are those of a mind fully attentive, and attentive as related to observation out of a non-dualistic state of mind, a state of mind where there is no observer observing, are those which voyage in relationship to real discovery, discovery of what really *is* and not of what was or what might, should, or will be. To have the partially seeing eyes of a rational philosopher, for example, or of a systematic theologian, eyes which might be said to be left-brain rather than right-brain eyes, or to have the partially seeing right-brain eyes of a romantic like Proust is not to have, so to speak, quiet mind "eyes" which can really discover things deeply significant, things which relate to the real, the unknown, the eternally alive. Partially seeing eyes cannot really see and discover the new, the fresh, the unknown, the immeasurable but only the old and dead, the already known and recognized, the measurable, the experienced. Such eyes might search for what is called "the essence of a human being,"[33] but what it will find as a result of any such search is merely some projection out of a partially functioning mind and never anything real. Such eyes never see anything in the land of the living but rather can only look about or move around in the house of the dead.

In his "Me Millennium" article, Luc Sante deals with the question of the so-called uniqueness of the "individual"[34] and says in effect that it is obvious that anyone of us is unique since the circumstances of birth of each one of us and the never-the-same features that are there when each of us is born make each of us unique. What this author could have made clear is that, of course, physically speaking, there are *real*, unique individuals, but that psychologically speaking there are not. So-called unique individuals, pschologically speaking, are not anything real but merely fashionings of a partially functioning mind. Such fashioning relates to a so-called knowing that tendencies, peculiarities, idiosyncrasies, and so on, make each one of us "an inimitable human being."[35] What is truly important, however, is to be self-knowing, that is to realize the true nature of the self, the so-called unique "human being,"[36] the truth about it, that the "you"[37] about which one speaks when there is talk about consciousness which relates to one's knowing one's own uniqueness is not at all a reference to anything real but rather to something which is merely mind-made. Any and every self is unreal, and, when it is "proclaiming" its "individuality"[38] as if it were something real, it is engaging in illusory, false-related activity. Moreover, any "kinship with others of the same micropersuasion"[39] which it might, likely will, seek to establish could or would be nothing more than one image, that which is unreal, relating to other mere images, equally unreal. Any relationship which is there then would not at all be a real one, not one out of love but rather merely out of thought.

There can be no real relationship with anyone or anything when one walks always in the tunnel of opposites, including that of collectivism versus individualism. To live, on the one hand, a life of conformity, "decentralized"[40] or not, to be a part of a "focus group ... which promotes adherence to a norm,"[41] or, on the other hand, to think that there is a real "you there,"[42] which, in reaction to societal conditioning influences, sands "down" its "edges," tailors, trims, and adapts itself, and goes "underground, meaning somewhere deep within"[43] is to live in relationship to what is not real. It is to live in relationship to the unreal fashioned by a busy partially functioning mind, and not in relationship to what is real, of the eternally, actively present moment, not in relationship to a truly meditative mind, which, in its deep passivity, understands fully and profoundly, which really lives rather than merely thinks, which allows for partial functioning on its part when such functioning is necessary and appropriate, but which, in its measureless silence, is otherwise profoundly wise and loving, deeply aware and alive.

Lauren Slater, at the conclusion of her "Me Millennium" article, tells her daughter to "make" her aging mother "into a myth" which portrays her as either a villain or a heroine.[44] What she is asking is that her daughter spend time in a house of illusion, a house that is not real but exists only in memory, in imagination. Myth is unreal, as unreal as anything else related to mere mind projection on the level of opposites, for example what is lovely to behold versus what is not. When this author suggests to her daughter that, as a parent ages, it is necessary for his or her child to "preserve the past,"[45] she is advising that this child move about in a house of the dead, of memories. One who lives only, or even largely, in the past, that is in relationship to memories, can "live on" but cannot "live largely"[46] as Slater would have her daughter do. Only when there is openness to the eternal, the living, that with which there can be communion when there is a truly meditative mind, can there be living in relationship to the whole, the new, the unknown, reality.

When there is a really meditative mind, there is awareness beyond the consciousness which relates to "long-privileged" groups which "value only those things that reflect themselves,"[47] and, on the other hand, which reflects the so-called "wider culture"[48] which, no matter how wide it is, is still limited. The awareness of anyone who is part of such a culture, because it is on a merely conscious or self-conscious level, is hedged in to a greater or lesser degree, but to some degree, to some measure, nonetheless.

To oppose the group, however, whether the smaller or the larger which is the dominant culture or the nation, for example, by turning against "stereotypes" so as to produce "your own records" as well as "distribute them," to "find a place for yourself on cable," or to "produce your own plays" and "publish your own books"[49] is still to stay on the level of a partially functioning mind, the level of opposites, of confusion and conflict, of small-mindedness, and, so long as mind is always and only of this level, there cannot be seeing without see-er, which is seeing the "genuinely new,"[50] reality, the living. So long as there is merely the look-er looking "into the mirror of our culture" there cannot be the seeing of "something genuinely new looking back."[51]

In light of the insight and understanding which are of a truly meditative mind, it can be said that the "me" is insignificant, but so is the group, the culture. True wisdom relates to this realization. It is illusion on the part of an individual to entertain a "secret dream about 'showing them,'" that is, showing society, one's so-called individuality, one's true self. It is illusion pure and simple to think that "when our real, best selves are revealed, we will be appreciated for what we really are."[52] There is no real self to be revealed or appreciated. "Good dancers" and "heroes"[53] are merely creations of a partially functioning mind, hence not anything real. Likewise unreal, because also merely mind-created, is the collectivity, the group, the external or exoteric opposite of the internal or esoteric. Beyond these opposites and all other opposites is the eternal, and it is quiet mind communion with the eternal which is highly significant and of deep worth.

In her article, entitled, "The Tricks Mirrors Play," Margo Jefferson speaks in obvious dualistic terms about "the duality of spiritual and secular life,"[54] about people "able to recognize themselves in a mirror, to be two things at once, the observer and the observed,"[55] but she does not point out that a truly meditative mind is aware of the falseness of such a dualism and that to think in terms of it is to think in narrow, limited terms and ultimately, therefore, not to understand fully or totally that which is living and of the living, of the new and the unknown, namely reality itself, life in all its dynamism and richness, that which is of the eternal or the always present rather than of the dead and gone – the static, the experienced, the known, that which can be recognized and projected from out of a partially functioning mind as something inert and abstract, that is, symbol, myth, concept, idea, notion, and so on. Merely casting "a glance backward and forward and to think about what we have been and what we want and want not to be"[56] is to be in the land of the dead.

With talk about "self-revelation, self-love and self-hatred, self-assurance and self-doubt,"[57] Jefferson makes the point that reflections like these which "we are capable of seeing keep multiplying," and that "as they do, this 'we' that we so comfortably talk about, you and I, cannot be taken for granted."[58] One of the points which could be made here is that there is a kind of observation which is without observer, which is to say, a level of attention, of awareness without consciousness, without self at all in any form being present, hence with no one of these reflections mentioned by Jefferson being there. When there is such observation, there is attention which is of a mind which lovingly understands rather than understands in a merely abstract, verbal way, and the insight into living truth and the wisdom which relate to such understanding. Given this understanding, there is possible a way of living totally beyond any "dire consequences for individuals,"[59] as well as beyond any so-called "distorted reflections"[60] of groups, cultures, nations. Such a way of living has nothing at all to do with "self-flattering images of sacrifice and grandeur" which someone might say "can serve as a source of self-esteem and pride – or as a rationale for crimes against humanity."[61] It is a way of living, not in relationship to opposites, for example, "the values of Western civilization," examples of which we see in someone like John Paul II and which are fundamentally no different from those of the Buddhist Promoting Foundation, "versus those of multiculturalism"[62] but a way of living in relation to the whole beyond all opposites, that which is living rather than dead, that which is of the eternal moment or the active present rather than of time with its past and future, and present as merely a link between past and future.

To long to be a child of anyone, for example of "Descartes" or "Christ,"[63] whether of someone who actually lived or lives now, or someone or something merely imagined, that is, simply mind-created, mind-projected, is to be a slave to that someone or something, with a narrowed-down, superficially functioning mind. It is not to have a mind totally free, completely uninfluenced or unconditioned, is not to have a mind alone. Only a mind alone can directly relate to the depths of that which is living, the eternally now, the new, the unknown. To have a narrowed-down, superficially functioning mind is to have a mind which creates the unreal, like an "invisible, untouchable soul,"[64] what is merely a result of speculation and/or imagination, or a mind which fashions "a new kind of mechanistic monotheism"[65] which has to do with a joining of "what is godly and what is flesh."[66] To engage the mind in order to effect a oneness out of partially functioning mind created fragments or opposites, for example the so-called spiritual and the so-called material, is to engage the mind in futile activity, activity which cannot lead to an understanding of living truth but merely to what is false or illusory.

As it is now, confusion and conflict abound, and there is little, if any, realization that deep understanding and a deeply meaningful way of living in relation to such understanding are possible. Andrew Cherlin points to such confusion and conflict when he reports that "Americans like to lament their individualistic bent, all the while pursuing it," and when he says that "the most fundamental quality of the American sense of self ... is neither individualism nor commitment but rather ... continuing ambivalence" as this self steers "a life course between them."[67] As this course is steered, and so long as there is merely the observer observing this course being steered, there is no realization that profundity of understanding and living are not discovered when there is only a mind which moves in relationship to opposites but rather when opposites are negated and the mind is totally free beyond them. Such freedom is the negation of what *is*, including confusion and conflict created by a partially functioning mind. Only when there is this freedom is there the seeing of reality, rather than something which is partially functioning mind fashioned, hence that which is unreal, even false or illusory, rather than mere speculation related to an "expectation of the moment when" one's "life would become whole."[68] Only a mind never aware of total freedom on the level of a truly meditative mind would ask whether or not the only possible "freedom" there is has to do with rearrangement of "reality more or less to our liking."[69]

Confusion and conflict dissipate when there is total attention paid to them, that is, when there is observerless observation and an accompanying understanding of such confusion and conflict. That undissipated confusion and conflict are there in our thinking and living is pointed out when Cherlin says that "to judge by new evidence, the cycle has swung back again to individualism, indeed swung past self all the way to selfish."[70] If, however, "self" here is intended to refer to something real, then this also is confusion for self, no matter what its form is – higher, lower, or something else – is not anything real, but indeed, as David Samuels says, merely "a fictional construction of the mind."[71] People are confused when they "bemoan the self-centeredness they see around them" while "at the same time, they express starkly individualistic views," when "they strongly identify with personal responsibility, self-sufficiency, and self-expression."[72] There is confusion when there is "expressive individualism" related to "emotional gratification, self-help, getting in touch with feelings, expressing personal needs" related to a "going to the forest ... to find" a so-called but partially functioning mind-made "inner essence."[73]

Belief that there is an inner essence, a soul or higher Self as different from the ego, the lower self, is an age-old false or illusory notion or idea. Long has there been belief in the "contemplative individual" or "self"[74] as something real. Contemplation, however, is merely an activity of a partially functioning mind, one of many activities which involve mere mind projection of what is supposedly real, but which is merely something imagined, abstract, unreal. George Johnson talks about the historical movement over time away from contemplation in relation to the external or so-called objective world to the realm of the so-called subjective, that is to "the land of the inner self."[75] What we have not at all come to realize is what is clear on the level of the truly meditative mind, namely, that there is no inner self as anything real. We have not come to the realization that there can be profound awareness which is other than consciousness beyond all forms fashioned by a partially functioning mind, be they exoteric or esoteric, and beyond the thinking of both materialists who feel certain that a scientific explanation of the nature of consciousness is possible and that of the so-called "New Mysterians" who "believe that ... we" might "never" comprehend "consciousness" since it "is so far outside the domain of conventional science."[76] Such awareness is not at all related to partially functioning mind fashionings of any kind.

Only deep understanding on the level of a meditative mind can bring an end to "the conflict between the individual and society."[77] Such understanding is prerequisite to a way of living related to a mind which is really silent, meditative, loving, but which properly, scientifically functions whenever such functioning is necessary and appropriate. Were the minds of many, or even all, individuals – individuals, however, not in the ordinary sense of opposed to others, to society, but in the sense of minds which are whole, undivided, indivisible, not bifurcated or fragmented – related to loving intelligence and understanding, to true meditation, then society would likewise be related to such understanding rather than to the partially functioning mind understanding and talk which have characterized all societies down to the present day, and which characterize as well smaller communities or groups within the larger societies that are ours today.

Daniel Menaker calls Werner Heisenberg "The Radical Thinker."[78] Heisenberg was indeed such a thinker if radical is taken in the sense of opposed to the norm, the standard, the usual, or the ordinary, but from the level of the truly meditative mind it can be said that all thinkers and all thinking relate not to what is totally radical, ultimately revolutionary, that is, the awakening of the profound intelligence of the truly silent, meditative, religious mind. Such profundity of intelligence and the state of this truly silent mind are requisite if there is to be a full, complete answer to the question of what it means to be human. If and when the minds of Heisenberg and Einstein "knew, or at least intuited" that "to be human" means "to have a conscious, cogitating self and to try to understand that self's relationship to the universe,"[79] theirs was a narrow understanding of what it is to be human.

After Heisenberg formulated the uncertainty principle and suggested that "we can never really know the world ... because our very efforts to do so change and in a way corrupt the world we are trying to know," because "the conscious human self" is "an automatic interferer, a changer, a polluter of reality," because "the human self ... by its nature and definition radically isolated from its context ... in trying to overcome this isolation ... must tincture and perhaps even taint whatever it finds,"[80] he should or could then have considered the possibility of awareness which is not opposite to but rather beyond conscious awareness and the possibility of mind being totally silent and in its complete silence in communion with the living, the new, the unknown, that is, reality, of mind beyond all the limitations which are there when it functions partially, as it does in relation to science, technology, and the practicalities of living. From the level of the truly meditative mind one can see, choicelessly and observerlessly, the limitations relative to awareness which is consciousness, that awareness which partially functioning minds, like that of Chopra, for example, might consider revolutionary, but which is and can only be, at best, reformatory since it is premised on false assumptions, namely, that all awareness is consciousness, that beyond ego is a real self, that all of the mind is partially functioning, and that the known, or what is related to the known, can reach, fathom, or encapsulate the unknown.

One might well be able to say that Julian James' theorizing in *The Origin of Consciousness in the Breakdown of the Bicameral Mind* – a book mentioned in the New York Times "Me Millennium" article by George Johnson[81] – relates to intellectual truth about how the partially functioning mind which is the self developed. Never in anything said by Johnson or James, however, is there any reference made to the significance or the importance of mind which functions totally, that is, the whole of the mind, mind which can function partially and which can also be really silent, mind not divided into right and left sides or hemispheres and not a mere synthesis or amalgamation of left and right brain hemispheres. The mind of man can be silent, alertly passive, but it can also function rationally, that is, sanely and objectively whenever such functioning is necessary and appropriate. The truly silent mind is profoundly aware because its awareness is not merely conscious awareness, and, because it is profoundly aware, it can understand fully. Its relationship is to that which is living and not to what is dead, not to theories, notions, ideas, or concepts, including any having to do with a supposed real self.

Related to so-called right and left hemispheres of the brain is Wilber's talk of stages of consciousness development, and also, by the way, Lawrence Kohlberg's talk of stages of moral development. Such talk perhaps clarifies aspects of the development of religion and of morality, but it says nothing about the reality of the truly religious mind and of the profound morality of such a mind.

What Harrison and De Mello in effect are doing in their works mentioned and considered earlier is they are trying to revert to a state of mind development which predates the breakdown of the bicameral mind, and this by using a partially functioning mind to react against Wilber's fourth stage of consciousness development, that is, the stage of the rational mind in order that there might be a recapturing of what is considered to be wisdom associated with the mythic, third stage of consciousness development. Rather than staying always and only on the level of the dualistic mind, mind which opposes and reacts, what Harrison and De Mello could have done, when it comes to matters relating to the significance of living, relating, loving, is talk about negating all stages of consciousness development, is talk about the mind stopping all its partial functioning which relates to consciousness so that there might come the understanding, insight, and wisdom of the truly meditative mind. Theirs could have been, this is to say, negation not only of Wilber's stage three of consciousness development which relates to the creation of images and religious myths and also of his stage four of such development, the stage on which, in the case of religion, there is an application of reason and logic to what are said to be truths couched in religious myths, but also negation of all so-called higher or highest stages of such development. It is a prizing of the application of reason to so-called insights and so-called truths related to religious myths which one sees, for example, in John Paul II's *Fides et Ratio* and in The Buddhist Promoting Foundation's *The Teaching of Buddha*.

There is a place for proper partial functioning of the mind which is not bicameral, mind which relates to Wilber's fourth stage of consciousness development, functioning in relationship to an "I" facing an "It," that is, a subject that knows facing an object known. Such an "I," a knowing subject, however, is never anything real, anymore than is the "It," the object which is known. The limited arena for such proper partial mind functioning is that of science, technology, and the practical aspects of living. Such an arena is that of the superficial and never of the profound to which alone a meditative mind relates. In the state of the truly meditative mind there is no personal "I," no real inner, higher or deeper self, that self which is very much emphasized whenever there is the prizing of either the bicameral partially functioning mind or of the mind which is non-bicameral but still partially functioning. It is prizing of the bicameral partially functioning mind which we see more and more of these days, for example in De Mello's *Awareness*.

As thinkers and writers strive for a "science of consciousness,"[82] it might be said, they, so to speak, have their blinders on for they never at all entertain the possibility of awareness as larger than consciousness and mind as more than merely partially functioning. Even if there is eventually a "science of consciousness" and thus dissipation of the so-called mystery which is consciousness, and even if skeptics like Colin McGinn and David Chalmers[83] are right when they talk about the seeming impossibility of a scientific understanding of the nature of consciousness,[84] it remains that beyond the limits of consciousness, and hence of knowledge, of the known, the measurable, in a word, science, is that silent awareness which is not consciousness. Only out of such awareness can come insight into and understanding of reality, the living, the eternal. Even if, eventually, science arrives at a complete theoretical explanation of all that is physical, a fully conscious awareness and even total knowledge of everything that physically exists,[85] all of this would not at all relate to a knowing of reality, mystery, that is, the no-thingness which, though it is not and cannot be known, not even by a truly meditative mind, is nevertheless communed with in the state of such a really meditative mind. Such communion is the meaning of life, meaning, however, which is living, related to the living or the eternal and not merely an intellectual or abstract meaning or so-called meaning associated with a partially functioning mind.

To conclude that all awareness is consciousness directed either outward to objects in the so-called external world or inward to what is called the inner realm of the spirit or of one's own uniquely individual musings is wrongly to absolutize consciousness. Any talk of an external or internal world, a world which is merely a projection out of a mind which functions only in relationship to consciousness, is merely related to partial or fragmentary mind functioning and precludes therefore that fullness of insight and understanding which are there when mind is beyond the limits of awareness which is consciousness. Related to these limits is science itself and its observer observation which entails separation of subject seeing from object seen. When, however, one sees that, as Johnson puts it, the methodology of science "breaks down,"[86] that the "subject is the object,"[87] then the moment is there for transcendence beyond the limits of consciousness, of method, of thinking, of time, and so on, but such a moment is not, so to speak, grasped by contemporary thinkers who are merely busy trying to solve the so-called mystery of consciousness. Even if scientists eventually do arrive at the point where they achieve a solution to this so-called mystery,[88] their minds will still be confused and in conflict if they have not gone beyond or transcended the limits we wrongly set to the mind, and all because we conflate all awareness to consciousness, prize only some fragment or part of the mind and not at all the whole mind, and deny or ignore the possibility of communion with the eternal beyond the temporal, the unknown beyond the known, the ineffable beyond the sensate, the immeasurable beyond the measurable, silence beyond science.

Krishnamurti rightly said that "the ending of knowledge is the beginning of wisdom."[89] Thus the ending of science and of the knowledge which relates to the doing of science is the beginning of wisdom, of the truly religious mind. Likewise, the ending of religion, of faith which in the case of theism is said to be related to knowledge and which in the case of fideism is said not to be so related, is the beginning of the truly religious mind.

All of these considerations lead one to the realization that, though we need science, we also need silence. Why we very much need silence is because science, whether or not there is any attempt to wed it to mysticism, or mysticism to it, cannot fathom the mystery which is living, namely, the unknown, the immeasurable, that which is ultimate reality. Science can never be the whole or the complete for science is thought – theoretical and applied – and thought cannot be other than limited. There is real dualism of time and the timeless, and science has to do with time, the limited, and meditation with the timeless, the unlimited. To realize such dualism is transcendence of both positivism and theism. Positivism was/is right to limit reason or knowledge as it did/does, but wrong to neglect, ignore, or deny the transcendent. Theism was/is right as regards its concern for the transcendent but wrong as regards its neglect of the proper limits of reason, of knowledge. The truly meditative mind sees the proper limits of reason or of knowledge while at the same time realizing the possibility of communion with the transcendent, that is, the really transcendent or the living eternal, when there is the state of the truly meditative mind.

It might be said that psychology in its so-called transpersonal forms not only does not have eternal value but has no value at all since it relates to the false, the illusory, since these forms are neither science nor silence, and the same is true of religion as well, whether deriving from conventional, left-brain or from right-brain, perhaps mystical, thinking, which likewise is neither science nor silence. What has eternal value, and so what we truly need today, it might be said, is the truly meditative mind for it alone can commune with the eternal. What we certainly do not need, particularly, are the illusions associated with so-called New Age religion or spirituality. Such religion or spirituality, like any mainline religion and its associated spirituality, is based on a failure properly to realize the difference between time and the timeless, to realize that there can be the timeless only when there is an end to time. Such a failure is clearly indicated when one reads some of the promotional literature created by promoters of such religion or spirituality, when one hears their talk about conscious creations which will attune people with spirit, meditation made easy, intuition technology, inner dynamics, a holistic practice of some profession like law, and so on.

Bede Griffiths wrote things which are very much like or similar to some of the things said by Chopra, Wilber, and Nisker, and by advocates of so-called spirituality, either so-called New Age or in its more traditional mainline religious forms. Griffiths, for example, talked about the limitations of "living from one half of our soul, from the conscious, rational level," the need "to discover the other half, the unconscious, intuitive dimension," and about his own desire "to experience ... the marriage of these two dimensions of human existence, the rational and intuitive, the conscious and the unconscious, the masculine and feminine," that is "the marriage of East and West."[90]

The same vision of a union of science and so-called ancient right hemisphere wisdom which we see in Nisker, Chopra, and Wilber is there when Griffiths talks of a "new consciousness" relating to "a new age of spiritual wisdom ... coming to birth," the way to which, he says, "science itself has prepared," when he talks of a "new vision of the world, which is also the vision of the ancient seers."[91] All of what Nisker, Chopra, Wilber, and Griffiths say, and others like Teilhard de Chardin say or have said along these lines is premised on two partially functioning mind derived false assumptions, namely, that all awareness is consciousness and that there is a real self, a soul. Griffiths, for example, makes consciousness too absolute when he talks about the "pure identity of Being and Consciousness,"[92] and when he says that the soul is "a reflection of the divine light in us."[93] The first of these two, just mentioned false assumptions relates to a failure to realize the true dualism of time and the timeless, the temporal and the eternal, and to the failure to see the difference between a partially functioning mind and one which is completely silent, really meditative.

When there is no religion at all, whether traditional or so-called New Age, or some other, but rather only science and silence, when there is not just science but silence as well, then there is not just the superficial, however necessary the superficial sometimes is, but there is also the deep, the profound. When there is science and silence, but no religion, then there is the truly religious mind, that mind which is not a partially functioning mind transformed into a so-called meditative one.

It might be said that what our world truly needs, but is sorely lacking, is meditative mind awareness. This is clear, for example, when one sees out of a really meditative mind all of the opposing groups and opposing beliefs which are prevalent in the world today. There is merely opposition there when one sees, for example, that when troubles or adversities of life come their way, some people reject the way of religion and its related belief or faith in God while others take "a leap of faith,"[94] and, as they counter despair with hope, they use these very troubles or adversities as an opportunity to deepen their so-called faith or trust in God. Mind in a truly meditative state realizes that despair and hope are merely opposites, and confusion is always there when, as regards living and relating, there is merely movement in the hallway of opposites. Such a mind is aware that opposition and confusion are there in what Ron Graham noted and talked about when he wrote that, though in organized religions there are "good people"[95] who act charitably, there is also the ritualization of "wisdom," bureaucratic "structures," and much "politicized" activity.[96] When, however, there is awareness without consciousness, a truly meditative and not just a partially functioning mind, not only is there no confusion and no conflict but there is an understanding of and communion with ultimate goodness, wisdom, and truth and not merely some limited, partially functioning mind related so-called good or so-called wisdom associated with a self supposed to be something real. It is just such a self which, though it is merely a projection out of a partially functioning mind, is supposed to be related to ultimate reality, called spirit. It is not ultimate goodness, profound wisdom, and living truth which religious believers can find in the particular organized religion to which they belong.

Belief in such a so-called spirit causes people to say, for example, that life is more powerful than death and that spirit will ultimately triumph over flesh. Statements like these reflect a separation of life and death so that there is affirmation of the one and denial of the other, a setting up of the unreal dualism of spirit and matter, and the suggestion that person or soul, sometimes called spirit or said to be related to spirit is something real. It is precisely false dualisms of spirit and matter and of soul and ego which are there when Graham, out of a partially functioning mind, talks about so-called spiritual greats or giants who withdrew from the world and who, having passed "through a dark night of the soul,"[97] broke their lower selves and triumphed over fear. If anyone of the so-called masters of the spirit believed in the soul as something real, or at least so long as he or she believed this, his or her mind too was ever functioning partially and was not truly meditative.

Graham is right when he says that people have made an "abstraction" or "entertainment" of death.[98] When, however, he goes on to suggest that we need greater awareness of the reality of death, he assumes the presence of the self being aware, but any awareness with self there is not deep awareness on the level of the meditative mind. Only such deep awareness can be a full and complete answer to questions about the importance and significance of psychological living and dying. A lack of full and complete awareness is there when there is talk, like that of Graham, about "spiritual values" opposing "material" values.[99] Values like "love and compassion"[100] are real only if there is no false dualism of spirit and matter, and no psychological self there supposedly being loving and compassionate. Though the psychological self can be conscious of the brevity of our time in the world, and as a result of its conscious awareness can make changes as regards its "priorities" and "values,"[101] it cannot become profoundly wise. Because any process of becoming so-called "wiser and kinder"[102] in which it might engage itself is partially functioning mind related, and has nothing at all to do with what is real wisdom and real love, is perhaps why, after an event like that of September 11, 2001, it again concentrates on "getting and spending,"[103] and to rekindling its belief that it is "invincible," and "immune" to "fate."[104]

It is not a truly meditative mind which is there when Graham talks about constant awareness of death,[105] and when he talks about the prayers and "good works" of those who are called "saints."[106] A supposedly real self is there for Graham as much as it was there for the saints who prayed and did so-called good works. When Graham speaks about right living and right dying, both with calmness of "mind," courage in the "spirit," and love in the "heart,"[107] what one can say is implied by him is a supposedly real self being calm, courageous, and loving rather than agitated, afraid, and not loving. When he talks, moreover, about living right now and death as something which follows later, he unduly separates psychological living and dying. The unreality or the falseness of such a separation is seen by mind in a truly meditative state. When there is real psychological living and dying, there is no separation at all. Then the living is the dying, and the dying is the living, moment to moment, day to day. When there is real living, then there is dying in the moment, eternally so, to all mind conditionings, influences which are there when there is only and always a partially functioning mind. Only when there is the totally unconditioned mind is there mind which is pure, uncontaminated, totally alone, that is, mind which is whole, integral, fully understanding and profoundly wise, not fragmented, not in confusion and conflict. Only then, this is to say, is there mind fully alive, fully living. Only then is there real, total freedom of the mind and not merely the limited, indeed illusory, freedom about which Graham speaks when he talks about an implied real self exercising its "freedom" to live wisely in this world.[108] When self is there, there is no real freedom, no real love, no depth of understanding and wisdom, hence no living of a life which is profoundly meaningful, joyful, and peaceful.

Limited science and unlimited silence are the two appropriate answers to our two important questions. When there is silence and no religion at all in the usual sense, then there is the truly religious mind, then there is religion as "the total way of life" as well as "the understanding of truth, which is not a projection of the mind"[109] rather than as merely organized faith or religious belief and activities related to such faith or belief. When the value but also the limits of science are there, there is no attempt by science to fathom living mystery, and, when there is the truly meditative mind, mind has stopped all its partial functioning in relationship to living and relating. Let there be intellect and its intelligence, and let there be as well the loving intelligence of the truly meditative mind. Let there also be the arts, but it is important to understand that the arts and what artists produce are mere creations of partially functioning minds, of what might be said to be mostly right rather than left-brain partial mind functioning, and that these are always and necessarily merely time-related, that is, that they cannot be *the* way, or a better way of "knowing" the unknown, of capturing something of the unknown, the living, the eternal. The unknown cannot be known in any way; it cannot be encapsulated or enveloped in any fashioning by or as a result of effort on the part of a partially functioning mind, be that left or right-brain related.

Only when there is the loving intelligence of the meditative mind as well as science and art both properly understood and practiced can there be the living of lives which are fully human. Fully human living, it might be said, is living filled with meaning, with eternal love, joy, peace, and order, and not living related merely to time and a partially functioning mind, merely to a mind functioning rationally rather than irrationally, that is appropriately and necessarily in relationship to time and time-related things, including science and all of its intellectually meaningful creations, that is, its theories and concepts, as well as to the application of its findings which make for ease of living in relationship to time.

REFERENCES

Introduction: The Two Questions

1. J. Richard Wingerter, *Beyond Metaphysics Revisited: Krishnamurti and Western Philosophy* (Lanham, MD: University Press of America, 2002).
2. J. Krishnamurti, *Commentaries on Living*, edited by D. Rajagopal (Wheaton, IL: Quest Books, 1967), vol. 2, p. 93.
3. Ibid., vol. 2, p. 90.
4. Ibid., vol. 1, p. 180.
5. Ibid., vol. 3, p. 289.
6. Ibid., vol. 1, p. 244.
7. Ibid., vol. 3, p. 7.

Chapter One: Western Religion, Science, and the Meditative Mind

1. John Paul II, "Fides et Ratio," *Origins*, vol. 28, no. 19; October 22, 1998.
2. Anthony De Mello, *Awareness* (New York: Image Books, 1992).

   Specific Quotes from John Paul II, "Fides et Ratio," *Origins*, vol. 28, no. 19, 1998, listed by Chapter Section:

   Chapter Section 1.0:
   Cover page, p. 327
   Chapter Section 1.0 – by Commentator:
3. pp. 338-339
4. p. 328

5. p. 338
   Chapter Section 1.1:
   p. 338
   Chapter Section 1.1 – by Commentator:
6. p. 321
7. p. 319
8. pp. 319-320
9. Ibid.
10. p. 342
11. p. 323
12. p. 319
13. p. 321
14. pp. 319, 338
15. pp. 335, 343
16. p. 319
17. pp. 321, 345
    Chapter Section 1.2:
    p. 340
    Chapter Section 1.3:
    Cover page
    Chapter Section 1.3 – by Commentator:
18. p. 339
    Chapter Section 1.4:
    p. 337
    Chapter Section 1.4 – by Commentator:
19. p. 336
    Chapter Section 1.5:
    Cover page
    Chapter Section 1.5 – by Commentator:
20. p. 322
21. Ibid.
22. p. 320
23. Ibid.
24. pp. 323, 339, 341
    Chapter Section 1.6:
    p. 345
    Chapter Section 1.7:
    p. 325
    Chapter Section 1.7 – by Commentator:

25. p. 322
Chapter Section 1.8:
   p. 319
Chapter Section 1.9:
   p. 340
Chapter Section 1.9 – by Commentator:
26. p. 342
27. pp. 342, 344
Chapter Section 1.10:
   p. 325
Chapter Section 1.11:
   p. 321
Chapter Section 1.11 – by Commentator:
28. Ibid.
29. p. 337
30. p. 339
31. Ibid.
32. p. 340
33. p, 341
34. Ibid.

Specific quotes from Anthony De Mello, *Awareness* (New York: Image Books, 1992), listed by Chapter Section:

Chapter Section 1.12:
   pp. 175-176
Chapter Section 1.13:
   pp. 121, 123
Chapter Section 1.13 – by Commentator:
35. pp. 78-83
36. pp. 78-80
Chapter Section 1.14:
   p. 113
Chapter Section 1.14 – by Commentator:
37. p. 114
38. p. 59
39. Ibid.

40. p. 73
41. pp. 80-81
42. p. 87
43. p. 121
44. p. 126
45. Ibid.
46. p. 127
47. pp. 134, 136
48. p. 184
49. Ibid.

Chapter Section 1.15:
    p. 176
Chapter Section 1.16:
    p. 172
Chapter Section 1.16 – by Commentator:
50. pp. 125-126
51. pp. 116-117
52. p. 115
53. p. 107
54. Ibid.

Chapter Section 1.17:
    pp. 177, 132
Chapter Section 1.18:
    p. 152
Chapter Section 1.19:
    p. 96
Chapter Section 1.20:
    p. 174
Chapter Section 1.20 – by Commentator:
55. pp. 140-141
56. Ibid.
57. Ibid.
58. p. 173

Chapter Section 1.21:
    pp. 74, 176
Chapter Section 1.21 – by Commentator:
59. p. 141
60. Ibid.

Chapter Section 1.22:
   p. 170
Chapter Section 1.22 – by Commentator:
61.   p. 169
62.   Ibid.
63.   p. 150
Chapter Section 1.23:
   pp. 81-82
Chapter Section 1.24:
   p. 148
Chapter Section 1.25:
   p. 140
Chapter Section 1.26:
   p. 96
Chapter Section 1.27:
   p. 78
Chapter Section 1.28:
   p. 86
Chapter Section 1.29:
   p. 160
64. Anthony De Mello, *Walking on Water*, trans. Phillip Berryman (New York: The Crossroad Publishing Company, 1998), p. 45.
65.   Ibid., p. 25
66.   Ibid., p. 45
67.   Ibid., p. 47
68.   Ibid., p. 64
69.   Ibid., p. 19
70.   Ibid., pp.117-118
71.   Ibid., p. 118
72.   Ibid., p. 105
73.   Ibid., p. 106
74.   Ibid., p. 113

Chapter Two: Eastern Religion, Science, and the Meditative Mind

1. Buddhist Promoting Foundation, *The Teaching of Buddha* (Tokyo: Kosaido Printing Co. Ltd., 1980).
2. Steven Harrison, *Doing Nothing* (New York: Tarcher/Putnam, 1998).

Specific quotes from Buddhist Promoting Foundation, *The Teaching of Buddha* (Tokyo: Kosaido Printing Co. Ltd., 1980), listed by Chapter Section:

Chapter Section 2.0:
    pp. 80, 82
Chapter Section 2.0 – by Commentator:
3.   pp. 196, 414
Chapter Section 2.1:
    p. 142
Chapter Section 2.1 – by Commentator:
4.   pp. 138, 140
5.   p. 138
6.   Ibid.
7.   pp. 134, 136
8.   pp. 134, 136
9.   pp. 132, 134
Chapter Section 2.2:
    p. 136
Chapter Section 2.3:
    p. 140
Chapter Section 2.3 – by Commentator:
10.   p. 70
11.   Ibid.
12.   pp. 416, 418
13.   p. 356
14.   Ibid.
Chapter Section 2.4:
    p. 308
Chapter Section 2.4 – by Commentator:
15.   pp. 356, 206
16.   pp. 206, 294
17.   pp. 36, 152
Chapter Section 2.5:
    p. 208
Chapter Section 2.5 – by Commentator:
18.   p. 126
19.   Ibid.
20.   p. 78

21. Ibid.
22. Ibid.
23. Ibid.
24. p. 76

Chapter Section 2.6:
p. 376

Chapter Section 2.6 – by Commentator:
25. p. 286
26. p. 198
27. Ibid.
28. pp. 196, 198
29. Ibid.

Chapter Section 2.7:
p. 40

Chapter Section 2.7 – by Commentator:
30. pp. 40, 202, 204
31. pp. 202, 204

Chapter Section 2.8:
p. 154

Chapter Section 2.9:
p. 368

Chapter Section 2.9 – by Commentator:
32. p. 146
33. p. 378
34. Ibid.
35. p. 374
36. p. 300
37. Ibid.
38. p. 308
39. Ibid.
40. Ibid.
41. p. 302
42. p. 164
43. p. 198
44. pp. 206, 208

Chapter Section 2.10:
pp. 280, 282

Chapter Section 2.10 – by Commentator:
45. p. 378

46. Ibid.
47. p. 224
48. p. 166
   Chapter Section 2.11:
   p. 102
   Chapter Section 2.11 – by Commentator:
49. p. 130
50. p. 44
51. Ibid.
52. Evelyne Blau, *Krishnamurti 100 Years* (New York: Stewart, Tabori, and Chang, 1995), p. 226.
53. Ibid.
54. Ibid., p. 225
55. Ibid.
56. Ibid.
57. Ibid.
58. Ibid.

   Specific Quotes from Stephen Harrison, *Doing Nothing* (NewYork: Tarcher/Putnam), 1998, listed by Chapter Section:

   Chapter Section 2.12:
   Book Title
   Chapter Section 2.12 – by Commentator:
59. Book Jacket
   Chapter Section 2.13:
   Book Subtitle
   Chapter Section 2.13 – by Commentator:
60. Book Jacket
   Chapter Section 2.14:
   pp. 59-60
   Chapter Section 2.14 – by Commentator:
61. pp. 60-61
   Chapter Section 2.15:
   pp. 97, 100
   Chapter Section 2.16:
   p. 47
   Chapter Section 2.16 – by Commentator:

62. p. 16
63. Ibid.
64. p. 16
65. p. 25
Chapter Section 2.17:
  p. 2
Chapter Section 2.18:
  p. 46
Chapter Section 2.19:
  p. 40
Chapter Section 2.20:
  p. 64
Chapter Section 2.21:
  p. 53
Chapter Section 2.22:
  p. 52
Chapter Section 2.23:
  p. 26
Chapter Section 2.24:
  p. 29
Chapter Section 2.25:
  pp. 22-23
Chapter Section 2.26:
  p. 67
Chapter Section 2.27:
  pp. 9-10
Chapter Section 2.28:
  p. 35
Chapter Section 2.29:
  pp. 118-120
Chapter Section 2.29 – by Commentator:
66. p. 21
67. J. Krishnamurti, *Commentaries on Living*, vol. 2, p. 93.

Chapter Three:
Psychology: Science or Meditation?

1. J. Krishnamurti, *Total Freedom: The Essential Krishnamurti* (San Francisco: Harper, 1996), p. xiii.

2. Deepak Chopra, *Ageless Body, Timeless Mind* (New York: Harmony Books, 1993).
3. Donald Rothberg and Sean Kelly (eds.), *Ken Wilber in Dialogue* (Wheaton, IL: Quest Books, 1998).
4. Wes Nisker, *Crazy Wisdom* (Berkeley, California: Ten Speed Press, 1990).
5. Ibid., p. xii

> Specific quotes from Deepak Chopra, *Ageless Body, Timeless Mind* (New York: Harmony Books, 1993), listed by Chapter Section:
>
> *Chapter Section 3.0:*
>   pp. 215-216
> *Chapter Section 3.1:*
>   p. 42
> *Chapter Section 3.2:*
>   p. 28
> *Chapter Section 3.3:*
>   p. 325
> *Chapter Section 3.4:*
>   p. 108
> *Chapter Section 3.5:*
>   p. 7
> *Chapter Section 3.6:*
>   p. 37
> *Chapter Section 3.7:*
>   pp. 169-170
> *Chapter Section 3.8:*
>   p. 292
> *Chapter Section 3.9:*
>   pp. 318-319
> *Chapter Section 3.10:*
>   p. 257
> *Chapter Section 3.11:*
>   p. 172
> *Chapter Section 3.12:*
>   p. 32

Chapter Section 3.13:
   p. 28
Chapter Section 3.14:
   p. 302
Chapter Section 3.15:
   p. 283
Chapter Section 3.16:
   p. 280
Chapter Section 3.17:
   p. 315

Specific quotes from Donald Rothberg and Sean Kelly (eds.), *Ken Wilber in Dialogue* (Wheaton, IL: Quest Books, 1998), listed by Chapter Section:

Chapter Section 3.18:
   pp. 335-336
Chapter Section 3.19:
   p. 332
Chapter Section 3.20:
   p. 357
Chapter Section 3.21:
   p. 338
Chapter Section 3.21 – by Commentator:
6.   p. 335
Chapter Section 3.22:
   p. 339
Chapter Section 3.23:
   p. 315
Chapter Section 3.24:
   pp. 333, 343
Chapter Section 3.25:
   p. 313
Chapter Section 3.26:
   p. 310
Chapter Section 3.27:
   p. 358
Chapter Section 3.28:
   p. 327

Chapter Section 3.29:
    p. 355
Chapter Section 3.29 – by Commentator:
7.   p. 356
8.   p. 355
9.   p. 356
10.   p. 354
Chapter Section 3.30:
    p. 355
Chapter Section 3.31:
    p. 352
Chapter Section 3.32:
    p. 352
Chapter Section 3.32 – by Commentator:
11.   p. 337
Chapter Section 3.33:
    pp. 350 - 351
Chapter Section 3.34:
    p. 352

    Specific quotes from Wes Nisker, *Crazy Wisdom* (Berkeley, California: Ten Speed Press, 1990), listed by Chapter Section:

Chapter Section 3.35:
    p. 5
Chapter Section 3.36:
    p. 7
Chapter Section 3.37:
    p. 11
Chapter Section 3.37 – by Commentator:
12.   p. 73
Chapter Section 3.38:
    p. 12
Chapter Section 3.39:
    p. 14
Chapter Section 3.40:
    p. 91

Chapter Section 3.41:
  p. 109
Chapter Section 3.42:
  p. 119
Chapter Section 3.43:
  p. 120
Chapter Section 3.44:
  p. 121
Chapter Section 3.45:
  p. 149
Chapter Section 3.46:
  p. 174
Chapter Section 3.47:
  p. 177
Chapter Section 3.48:
  p. 178
Chapter Section 3.49:
  p. 180
Chapter Section 3.50:
  pp. 196-197
Chapter Section 3.51:
  p. 202
Chapter Section 3.52:
  p. 208
Chapter Section 3.53:
  p. 209

Chapter Four: Science and Silence

References in this chapter numbered 1 to 88 are from articles in *The New York Times Magazine*, Section 6, October 17, 1999. Various articles cited are listed as follows.

"The Me Millennium," editor of *The New York Times Magazine*, p. 20.
"The Rorschach Chronicles," by Margaret Talbot, pp. 28, 32, 34, 38.
"On Language," by William Safire, pp. 40, 42.
"I'm O.K., You're Selfish," by Andrew J. Cherlin, pp. 44, 46, 48, 50.
"In a Chat Room You Can Be NE1," by Camille Sweeney, pp. 66, 68, 70.
"How 'I' Moved Heaven and Earth," by Richard Russo, pp. 85-89.

"Werner Heisenberg: The Radical Thinker," by Daniel Menaker, p. 96.
"The Tricks Mirrors Play," by Margo Jefferson, pp. 98-99.
"The Incubator of Dreams," by Barbara Grizzuti Harrison, pp. 100, 102, 104.
"Mapping the Millennium: Inside Out," Genetic Self Portrait by Gary Schneider; text by John Noble Wilford, pp. 107-113.
"Prozac Mother and Child," by Lauren Slater, pp. 115-116, 118.
"In the Age of Radical Selfishness," by David Samuels, pp. 120, 122, 124, 126, 152-154.
"Tags: A Thousand Years of Identification," by Stephen Mihm and David E. Brown, pp. 128-129.
"The Lost Art of Immortality," by Michael Kimmelman, pp. 130-131.
"Terra Incognita," by George Johnson, pp. 132-134.
"Be Different (Like Everyone Else!)," by Luc Sante, pp. 136-138, 140.
"Coddled Egos," by Molly O'Neill, pp. 147-148.

1. *The New York Times Magazine*, Oct. 17, 1999.
2. p. 48
3. pp. 124, 126
4. p. 70
5. p. 20
6. pp. 40, 42
7. p. 131
8. p. 86
9. p. 87
10. Ibid.
11. pp. 87-88
12. p. 20
13. p. 100
14. p. 128
15. p. 88
16. p. 104
17. p. 100
18. Ibid.
19. p. 104
20. pp. 100, 102
21. p. 128
22. p. 147
23. p. 131

24. p. 89
25. Ibid.
26. Ibid.
27. Ibid.
28. p. 32
29. p. 28
30. p. 32
31. p. 89
32. p. 109
33. Ibid.
34. p. 136
35. Ibid.
36. Ibid.
37. Ibid.
38. Ibid.
39. Ibid.
40. p. 140
41. Ibid.
42. Ibid.
43. Ibid.
44. p. 118
45. Ibid.
46. Ibid.
47. p. 99
48. Ibid.
49. Ibid.
50. Ibid.
51. Ibid.
52. p. 86
53. Ibid.
54. p. 98
55. Ibid.
56. Ibid.
57. Ibid.
58. Ibid.
59. Ibid.
60. Ibid.
61. Ibid.
62. p. 99

63. p. 116
64. Ibid.
65. Ibid.
66. Ibid.
67. p. 50
68. p. 154
69. p. 153
70. p. 44
71. p. 120
72. p. 44
73. Ibid.
74. p. 132
75. Ibid.
76. Ibid.
77. p. 88
78. p. 96
79. Ibid.
80. Ibid.
81. p. 134
82. p. 132
83. p. 134
84. Ibid.
85. Ibid.
86. Ibid.
87. Ibid.
88. Ibid.
89. J. Krishnamurti, *Commentaries on Living*, vol. 1, p. 233.
90. Bede Griffiths, *The Marriage of East and West* (Springfield, IL: Templegate, 1982), p. 8.
91. Ibid., pp. 27-29
92. Ibid., p. 35
93. Ibid., p. 76
94. Ron Graham, "Death's Gift to Life," *Maclean's Magazine*, Dec. 17, 2001, pp. 15-16.
95. Ibid.
96. Ibid.
97. Ibid.
98. Ibid.
99. Ibid.

100.	Ibid.
101.	Ibid.
102.	Ibid.
103.	Ibid.
104.	Ibid.
105.	Ibid.
106.	Ibid.
107.	Ibid.
108.	Ibid.
109.	J. Krishnamurti, *Commentaries on Living*, vol. 3, p. 92.

INDEX

Agape, 105-106
Anderson, Allan W., 89
Arnold, Matthew, 119
Atman, 76, 90, 113
Brown, David E., 121
Buddha, 57-58, 62-63, 67-68, 70-72, 75, 130
Buddhist, 57, 70, 74-75
Buddhist Promoting Foundation, 58-74, 87-88, 118, 126, 130
canon, 100
Chalmers, David, 131
Cherlin, Andrew, 118, 126-127
Chopra, Deepak, 90-98, 114-115, 129, 133
Christ, 12, 126
Christian faith, 27, 55, 120
crisis of meaning, 32, 34, 37
de Chardin, Teilhard, 133
De Mello, Anthony, 2, 37-56, 88, 129-30
Descartes, René, 126
Einstein, Albert, 96-97, 128
Eros, 105-106
Graham, Ron, 134-136
Griffiths, Bede, 133
Harrison, Steven, 57-58, 76-88, 129-130
Heisenberg, Werner, 128
hermeneutics, 34-35
Houston, Whitney, 119
immanence, 20
James, Julian, 129
Jefferson, Margo, 125
John Paul II, Pope, 2-30, 34, 36-37, 55, 118-120, 126, 130
Johnson, George, 127, 129, 131

Kant, Immanuel, 21, 35
Kohlberg, Lawrence, 129
Krishnamurti, J., vii, xi, 1, 47, 52, 57-58, 75, 87, 89, 91, 94, 104, 107, 117, 132
language analysis, 34-35
McGinn, Colin, 131
Menaker, Daniel, 128
Mihm, Stephen, 121
mystic(al; ism), 38, 40, 42-43, 45, 55-56, 77, 85-89, 96, 98, 106, 132
New Age Religion, 1, 17, 132
Nisker, Wes, 90, 106-115, 133
Occam, William of, xi
Rinpoche, S., Professor, 74-75
Rumi, 114
Russo, Richard, 119-120, 122
sacred writing(s)/books, 3, 19, 22, 33
Samuels, David, 118, 127
Sante, Luc, 123
scripture(s), 3, 5, 19-20, 28, 34
Seed, John, 113
Slater, Lauren, 124
Sufi, 114
Sweeney, Camille, 118
Talbot, Margaret, 121
Wilber, Ken, 90, 98-115, 129, 130, 133
Zen Buddhists, 114

ABOUT THE AUTHOR

J. Richard Wingerter is a retired teacher living in Calgary, Alberta, Canada. Prior to his retirement in 1997, he taught school in Saskatchewan and Alberta. In 1973, he authored two articles entitled "Pseudo-Existential Writings in Education" and "Not Who but What is Pseudo?" which were published in *Educational Theory*. His other books, both published by University Press of America, Lanham, Maryland, are *Beyond Metaphysics Revisited: Krishnamurti and Western Philosophy* (2002) and *Teaching, Learning, and the Meditative Mind* (2003).